FIRE
EQUIPMENT

FIRE
EQUIPMENT

Ed Hass

ThunderBay
P·R·E·S·S

This edition published in 1998 by
Thunder Bay Press
5880 Oberlin Drive, Suite 400
San Diego, California 92121
1-800-284-3580

http://www.admsweb.com

Produced by
PRC Publishing Ltd,
Kiln House, 210 New Kings Road, London SW6 4NZ

ISBN 1 57145 158 7
(or Library of Congress CIP data if available)

1 2 3 4 5 98 99 00 01 02

Printed and bound in China

Contents

Preface

Classic 1930 Ahrens-Fox Model J-S-2 piston pumper #1272 assigned to Engine 21 in Detroit, MI, has just finished a tough pumping stint at the two-alarm Fromm's fire on January 6, 1940. The alarm came from Fire Alarm Box 421. *Courtesy Detroit Fire Dept., Clarence Woodward*

This book is an illustrated history of American firefighting equipment past and present, how it was used and is used today, how it evolved over the years, and what equipment from the past is most prized by today's collectors. You will see the personal equipment that yesterday's and today's American firefighters carried, such as uniforms, helmets, turnout coats, boots, breathing apparatus, etc. You will also read about the types of equipment carried on various types of American firefighting vehicles of the past and today, from the traditional pumpers, ladder trucks, and incident-command (chief's) cars, to modern helicopters, water-bombing aircraft, fire boats, and airport crash-rescue tenders.

This book will also show that not only has firefighting equipment changed from earliest times to today, but the methods of procuring this equipment have also changed. Before the Civil War, concerned citizens bought firefighting tools and donated them to their communities. Today, City Purchasing Departments accept bids from established manufacturers, and then sign a contract with the lowest bidder whose product meets the community's specified needs.

Manufacturing methods have also changed, from hand-crafting firefighting tools as needed, to mass production of standard firefighting tools, and today's factories that use computers and robots to perform repetitive manufacturing tasks.

Even the way America's firefighters are summoned to a fire has changed dramatically in over 300 years of America's fire service, as you will discover as you read on.

Introduction

Before hand pumpers gained popularity in the late 1700s, American towns and cities relied on bucket brigades that passed leather buckets of water from the nearest river, lake, or horse trough, to the fire. This modern replica of a colonial leather fire bucket, in the author's collection, depicts an 1852 Hunneman handpumper, serial number 452, the "Papeete," which survives at Columbia, CA. *All photos, unless credited otherwise, are by Simon Clay*

Let's imagine that we had a video camera to record glimpses of America's fire service and fire engine industry at different points in our country's history. What similarities and differences might we see in different eras?

Colonial Era

Scene 1 of our video. The time is the late 1600s. The place is a small farming village in the Massachusetts colony. The village night watchman, a combination of firefighter, constable, and town crier, spots a wisp of smoke curling tauntingly from the thatched roof of a storage shed.

He runs into the Village Square, twirling the watchman's rattle that he hand-carved himself. This watchman's rattle is a simple device, consisting of a wooden handle, a wooden gear, and a pair of wooden reeds that strike the teeth of the gear as the watchman twirls it.

The near-deafening clatter of the watchman's rattle wakes the sleeping villagers. The heads of each household run outside, grabbing the hand-sewn leather fire buckets that a village ordinance requires them to keep on the outside of their homes. With buckets in hand, they follow the watchman back to the fire.

At the fire, the citizens form a "Bucket Brigade," a double line from a nearby creek to the fire. One line passes the filled buckets up, where their contents are tossed directly onto the fire. The second line passes empty buckets back down to the creek for refilling.

Their efforts prove futile, and the storage shed burns to the ground, leaving a smoldering pile of ashes. They continue to toss water onto the ashes, so at least the fire does not spread to the farmer's house and the rest of the village.

Satisfied that the fire is out and will not rekindle, each citizen grabs his bucket, on which his name is painted, carefully hangs it back on the exterior wall of his house, and tries to return to sleep, praying that the next fire does not strike his own property.

In some Colonial communities, citizens pooled their money to buy the town some ladders, and poles with hooks on the end for pulling-down burning roofs. Sometimes, these ladders and hooks were carried to fires on a hand-drawn cart, hence the origin of the "hook-and-ladder" truck.

In a few rare instances, citizens of a particularly prosperous community might have imported a crude, one-cylinder hand pumper from England. These simple machines had no wheels, but poles on both sides that allowed four men to carry the corners of the engine on their shoulders.

Young America

Fast-forward 150 years, to the bustling city of Boston in the early 1800s. America has gained its independence from England, and has become economically prosperous, exporting American crops such as cotton, corn, and tobacco, and importing goods from every corner of the world. Scurrying to unload cargo from ships newly-arrived from Europe and the Far East, dock workers do not at first notice wisps of smoke around the window of a wooden warehouse at the far end of the pier. But quickly enough, those wisps turn into dark, black billows of smoke lofting skyward.

The shipping company's hired watchman grabs his rattle from his office desk, and heads into town to summon help. The rattle looks similar to the one that the watchman of the1600s used, but it has only a single wooden reed. The unseen difference is that the watchman did not carve this rattle himself. Instead, it was made in a local factory that employs dozens of craftsmen to make a variety of wooden products for a retail market. His employer gave it to him as one of the tools of his trade.

The shipping company's watchman runs

through downtown Boston, twirling his rattle and shouting, "Fire at Pier 2! Fire at Pier 2!" He passes the office of a prominent attorney, who stops in the middle of consultation with a client to grab his coat and hat and run to the little shed that houses Boston's volunteer fire company number 13.

A butcher, chopping meat with his cleaver for a waiting customer, hears the cry of "Fire!", quickly unties his bloody apron, and dashes after the lawyer. A shopkeeper, tallying his day's receipts, puts down his quill pen and also dashes into the street.

They all arrive at the sturdy little brick shed with "Fire Company Number 13" painted above the wooden doors. There, the lawyer, the butcher, the shopkeeper, and several other volunteer firemen from all walks of life, grab the thick, strong hemp rope attached to the front of their fire engine, and pull it behind them to the blazing pier. Unlike the crude hand pumpers of 150 years earlier, this one is mounted on two axles and four strong, spoked wooden wheels.

La Selva Beach, CA, built this replica of a 1678 hand pumper, illustrating the crude simplicity of America's earliest fire engines. Four citizens carried the small engine to fires on their shoulders, but the replica is hauled around on a modern child's wagon. This replica pumps water alongside newer, larger hand pumpers at many California firefighters' muster competitions. *Ed Hass*

Early Philadelphia fireman and fire engine. *Archive Photos*

9

The fire engine is named the "President Adams," and bears the hand-painted likeness of John Adams on both sides. Boston's own William C. Hunneman, a former apprentice of Paul Revere, built this engine.

In early America, fire engines were pulled to fires by hand, and the pumps were also hand-operated. In colonial days, they were made in England by Newsham or Mason, and brought over by ship, the expense usually paid by one or more generous local businessmen.

Later, some of New England's finest cabinetmakers, silversmiths, and other tradesmen would build fire engines for local communities as a sideline. Again, these engines were usually purchased and donated by the community's wealthier citizens.

A few of these early American craftsmen were so successful in their fire engine sideline that they decided to concentrate on building hand-pumped fire engines. The most prominent and successful of these was William C. Hunneman of Boston, a former apprentice to Paul Revere. Between 1792 and 1881, the company that he started built over 700 hand pumpers, not only for cities all over the U.S. from Maine to California, but throughout the western hemisphere, from Canada to Chile. The "President Adams" of Boston, in our imaginary video, was only the second engine that Hunneman had built.

Remarkably, several hundred Hunneman hand pumpers survive today, some after serving fire departments in many cities and many states. They survive in the hands of fire departments, museums, and private collectors. Many are on public display; others gather dust in barns and hay lofts. A few survived until fairly recent times, only to be lost to tragic fires or, still worse, scrapping by those who should have known better.

In the early 1800s, another popular builder of hand pumpers was James Smith of New York City.

Arriving at the burning dockside warehouse, some of the volunteer firemen take down the buckets hung along both sides of the fire engine, scoop them into Boston Bay, and dump the seawater into the engine's large wooden tub. The volunteer firemen know that the salt might corrode the metal pistons of the engine's simple pump, but this is the only nearby water source, and they

can always wash the engine out with fresh water back at their firehouse.

The fire company's leather buckets are little changed from those of 150 years earlier. But instead of hand-sewn seams, the bucket seams are reinforced with copper rivets. The buckets no longer bear the names of individual owners, but the name "Fire Company Number 13" and the image of President John Adams.

The hand pumper has a wooden pole on the front, and another at the rear. These poles, called "brakes," control the up-and-down motion of the two pump pistons. Once the engine's tub is full, the fire company's foreman shouts through an ornately-carved brass speaking trumpet to "man those brakes."

Five members of Fire Company 13 line up along the front brake, and five more along the back brake. They place their hands, including their thumbs, atop the poles. As one team pushes its pole down, the facing team's pole swings upward, in a see-saw fashion. These experienced volunteers know never to hook their thumbs under the brake poles, as the rapid up-and-down motion could easily break them.

While these men are busy pumping the brakes, other volunteer firemen continue to scoop water out of the bay, and empty their buckets into the engine's wooden tub, so the pump has a constant supply of water to squirt onto the fire. And the fire company

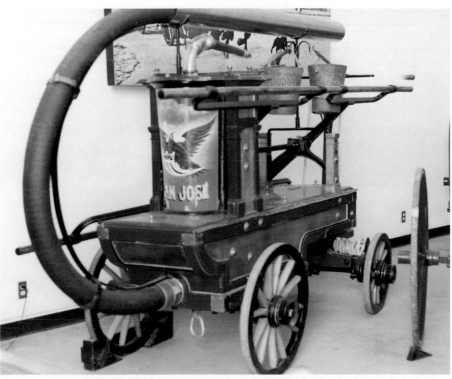

Left: Colonial-era hand pumper, the "Bolton Quickstep," at New York City Fire Museum (278 Spring Street). David Wheeler of Berlin, MA, built this engine in 1765. On the wall in the background are axes, hooks, and other typical early American firefighting tools of similar vintage to this engine.

Below: James Smith of New York City built this hand pumper for one of New York's volunteer fire companies in 1808. It later saw service in San Jose, CA, and is now displayed at the San Jose Historical Museum. *Ed Hass*

FIRE EQUIPMENT

foreman is directing it all, shouting commands through his speaking trumpet.

The ten men on the brakes soon tire from the vigorous exercise of working the brakes, and another crew of ten takes over while the first crew rests. A typical fire company of the era had 50 members, so there was always fresh muscle to "man those brakes!"

The water from the fire engine exits through a 25ft long leather fire hose, whose seams are reinforced with copper rivets. At the far end of the hose, a cast-brass, open-tip nozzle, 1-1.5in in diameter, directs the stream into the blazing warehouse.

Early hand pumpers were sometimes equipped with a copper or brass nozzle on top, to squirt water into a burning building. This required placing the wooden fire engine, and its volunteer operators, dangerously close to the flames and smoke. But by our early-1800s scene, leather fire hose, using copper rivets to reinforce its seams, permitted placing the fire engine and the firefighters at a safer distance from the fire.

After about an hour of vigorous pumping, the firemen have pushed back the flames and smoke that were streaming through the windows and roof of the warehouse when they arrived. The contents, only an hour earlier destined for Europe, are now burned beyond recognition. The roof has burned away, and the floor is covered with water and ash. But the walls are still standing, and the existing structure can be rebuilt.

The foreman shouts his final order through his trumpet: "Pick up!" The volunteer firemen scrub the grime off their leather fire hose, carefully fold it back onto the top deck of their engine, and hang their buckets back along the sides of their engine. They then grab the front drag-rope, and haul the Hunneman hand pumper back to the Number 13 fire station. There, they thoroughly rinse every last trace of salt out of the pumper, before each returns to his law office, butcher shop, or general store.

Firefighting Becomes a Profession

Fast-forward another half century. Now it is a September afternoon in 1851, and the place is Cincinnati, Ohio. We are about to witness a relatively minor fire, but one that would dramatically and permanently change America's fire service and its fire engine industry. I described this event in my 1986 book, *The Dean of Steam Fire Engine Builders*:

"It started with loud and distressed cries of 'Fire! Fire!' Out of curiosity, the (local newspaper) reporter followed the source of the cries, hoping to watch the brave volunteer firemen in action, and perhaps even a chance to grab a bucket and help."

By this time, the days of using the watchman's rattle had drawn to a close. Although firefighters were still often summoned by shouts, as in this case, many communities hung the heavy iron tires of locomotive wheels from sturdy supports in the Village Square, and struck the iron tire with a sledgehammer in event of fire. Usually, the city government either bought the locomotive tire, or had it donated by a local railroad company. Cincinnati kept a huge drum in a tall downtown tower, and its beat often summoned the volunteer firemen.

My 1986 book continues:

"The blaze turned out to be in a huge old planing mill at the corner of Augusta and John streets. En route to the fire, the reporter saw bankers, bakers, city councilmen, and men and boys of all ages dash out of stores and houses to their nearest fire house. There, they grabbed a section of drag-rope, and hauled the little hand-operated pumpers over the cobbled streets as fast as their legs would carry them.

". . . Washington Fire Co. #1 raced northward from their engine house on Vine between Front and Columbia, and began attaching their brass-riveted leather suction hose to the wooden fire plug at Augusta and John streets. Just then, Western Hose Co. #3 arrived after a mad dash southward from their fire hall on Fifth Street between Mound and Carroll.

"The foreman from Western Hose Co. loudly proclaimed that he had arrived before Washington Co., and tried to prevent Washington from using the hydrant. The Washington foreman indignantly replied that he and his men had arrived ahead of Western, and therefore had the right to use the hydrant. Apparently, neither foreman

noticed that the hydrant had two connections, so that the two companies could share it."

I will digress here a moment to mention that city water mains of that day, where they existed at all, consisted of hollowed-out logs laid end-to-end beneath the city streets, and filled with stagnant water. At strategic locations, a hole was tapped into the logs, and the hole plugged with a wooden hydrant, or "fire plug." One or two openings in the plug allowed volunteer firemen to connect hoses from the hydrant to the suction inlets on their hand pumpers, drawing water from the water mains. Today's fire hydrants use basically the same principle, except metal pipes have replaced the hollowed logs, and the water is under pressure from a central water pumping station.

Now back to our story:

"The argument over use of the hydrant soon escalated from shouts to fists, and even to hurling bricks and bottles at each other. While all of this was going on, of course, the fire in the planing mill was completely ignored.

"As the spectators watched, the flames spread from one end of the mill to the other, leaping defiantly in the air, and they felt the intense heat on their faces. The firemen, absorbed as they were in their bloody brawl, seemed to neither see the glowing flames nor feel the rising temperature. The crowd

At the time of the 1851 firemen's riot, Washington Engine Company No. 1 of Cincinnati used this Hunneman hand pumper, now displayed in the Cincinnati Fire Museum. *Ed Hass*

Gamewell of Newton, MA, cornered more than 90 percent of the fire-alarm telegraph market. They made Box #5327 of Fresno, CA, in 1931. This alarm box is now in the author's collection.

hesitated to attempt fighting the fire themselves, fearing the firemen more than the fire. Indeed, as the violence escalated, many found themselves retreating ever further to maintain a safe distance from the brawl.

". . . It was a matter of utmost pride as to which fire company arrived on the scene first, and which was the first to throw a stream of water on the fire. To come in second evidently called one's masculinity seriously into question. Long and bitter rivalries (between two or more fire companies) often owed their origins to really trivial disputes, and then other fire companies formed permanent alliances by siding with one or the other party in such a feud.

"Since the planing mill fire was not being fought, two more volunteer companies were dispatched to the scene. One had a long-standing alliance with Washington, the other with Western. Soon, both companies also joined the fistfight, escalating it from a brawl into a full-scale riot. The unattended fire, meanwhile, marched relentlessly on, spreading from floor to floor. Additional fire companies were sent for, and soon ten Cincinnati fire companies — a total of over 300 firemen — were rioting, and not one drop of water had touched the fire.

"By sunset, the smoke was visible clear across the Ohio River in nearby Covington, Ky. Covington Fire Co. #1 ferried its hand pumper across the river in what started as a neighborly gesture of assistance. But when Covington learned that their friends at Washington #1 were being beaten-upon by several rival fire companies, they also joined the riot rather than fighting the fire.

"Cincinnati's mayor, Mark P. Taylor, attempted to end the riot, but his efforts had no effect. The riot did not end until dawn, when the firemen were too exhausted and injured to fight. By then, the planing mill had burned to its foundation, making the dispute over the hydrant a moot point.

". . . A week after the firemen's riot at Augusta and John streets there was a growing movement among Cincinnatians to do something about it; not only to punish the firemen, but to make sure that it never happens again."

Part of the problem, it was felt, was that Cincinnati had too many firemen, with 14 engine companies, 3 hose companies, and a hook-and-ladder company, each with 30 to 50 members. Five of those fire companies had been organized only that year. The nature of the hand-operated equipment, and the fatiguing workout that hand engines gave to their operators, necessitated such a large number of firemen per company.

". . . About a month later . . . a special meeting of the Cincinnati city council . . . was called, to discuss what to do about fire department rowdyism.

"The recent riot was on everyone's mind. But many had not forgotten that only a few months earlier, virtually every fire company in the city was involved in an even-larger riot, this one protesting the visit of a Catholic archbishop to the predominantly Protestant city of Cincinnati . . .

"As a result of the two riots that year, several city councilmen, and some of Cincinnati's prominent citizens, clamored for fire department reforms to end the rowdiness of the volunteers. That is why the special council meeting was called.

"Prior to the two riots that year, two Cincinnati councilmen had been pushing for fire department reform for several years. Their names were Jacob Wykoff Piatt and James H. Walker.

"Piatt, a Kentucky native, recounted how most of the older and public-spirited firemen had retired over the past few years, to be replaced by young ruffians who considered a fire merely an occasion for starting a riot. In his opinion, the engine houses had simply become a place for the gangs of political parasites to loaf, or as councilman Miles Greenwood put it, a 'nursery where the youth of this city are trained in vice, vulgarity, and debauchery.'

"The problem, as Piatt explained it, was that the firemen served voluntarily, and therefore could not be discharged for misconduct. But, he reasoned, place the firemen on a paid basis, with a good measure of military-type discipline thrown in, plus the threat of firing for misconduct, and the rowdiness would cease.

"Piatt recounted that the first time he and Walker had proposed replacing the volunteer fire department with a paid system, the council chambers were packed with firemen who noisily demonstrated their opposition to reform.

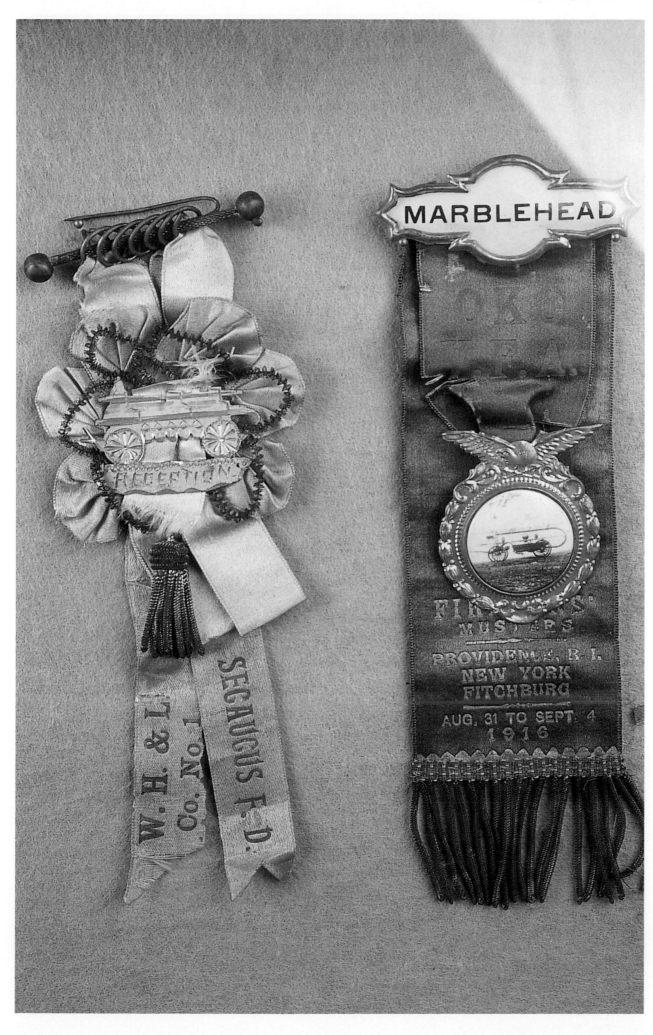

It is not clear for what occasion some brave firefighter was awarded the ribbon at left from Washington Hook & Ladder of Secaucus, NJ. But the ribbon at right was awarded to the firefighters of Marblehead, MA, for their victory in 1915 Firemen's Muster competitions at Providence, RI; New York, NY; and Fitchburg, MA. Both ribbons are displayed at the New York City Fire Museum.

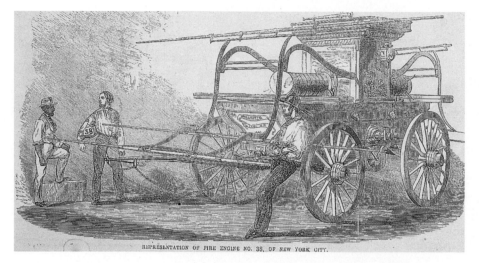

REPRESENTATION OF FIRE ENGINE NO. 38, OF NEW YORK CITY.

Drawing of New York City's Fire Engine No. 38, a 19th century shot. *Archive Photos*

"Outside of Piatt and Walker, the rest of the council had voted against reforms that day. America's volunteer firemen had amassed such tremendous political clout, Piatt said, that his fellow councilmen feared for their re-election, and in some cases for their lives.

"He and Walker had not given up, and had periodically re-introduced their fire department reform bill to the council. Each time, it was vetoed, but by an ever-decreasing majority. The firemen became so bitter about the reform efforts that, for his own protection, Piatt had to bring a guard of his Irish constituents to every council meeting. Once, an angry mob had even burned Piatt's effigy on his own front lawn, amidst hisses, howls, and yells.

"'How ironic,' Piatt gloated, 'that it had taken two terribly destructive riots before the city council joined me in seeking reforms.'

"At this meeting, the vote of the city council swung the other way: for the first time, the majority of councilmen supported Piatt and Walker's reform bill. Two vocal supporters were Miles Greenwood and Joseph S. Ross.

"Greenwood remarked that paying the existing firemen was not enough. The reason the firemen had succeeded in terrorizing Cincinnati and other cities for so many years was their sheer numbers: Cincinnati then had 18 fire companies, each with 30 to 50 members. Greenwood proposed that, in conjunction with establishing a paid fire department, Cincinnati needed to extinguish fires 'without the attendance of the large numbers of men required to operate the manual pumping engines.'

". . . Greenwood was renting space in his foundry to one Abel Shawk, who was experimenting with building a steam fire engine that would require fewer men to operate than the hand pumpers did."

Since 1829, there had been several unsuccessful attempts, in both England and America, to develop a practical steam-powered fire engine. Abel Shawk himself had entered an 1840 contest to develop a practical steam fire engine, but his entry did not win, and the winning machine saw only two months of fire duty before it was discarded.

Not deterred, in 1841, Shawk had acquired an 1819 Buchanan steam boiler from a mill in Johnstown, Ind. He began renting space (including use of a smith's forge) in Miles Greenwood's Eagle Iron Works in Cincinnati, dismantling the boiler to find ways to make it lighter, safer, and faster for practical use on a steam fire engine. He had completed his first prototype of an improved Buchanan boiler about 1849 or 1850, and was granted a patent on it on September 20, 1853.

"After the 1851 firemen's riot at Augusta and John streets, Cincinnati's city council became very receptive to Shawk's idea of building a steam fire engine. The councilmen realized that one steam fire engine could replace several hand pumpers and eliminate the need for hundreds of rowdy pump operators. So the council appropriated $1,000 to Abel Shawk to aid him in making practical tests of a prototype steam fire engine.")

Shawk used the pump from a retired

Advert for Holly's Patent Rotary Steam Fire Engine, c.1865. *Archive Photos*

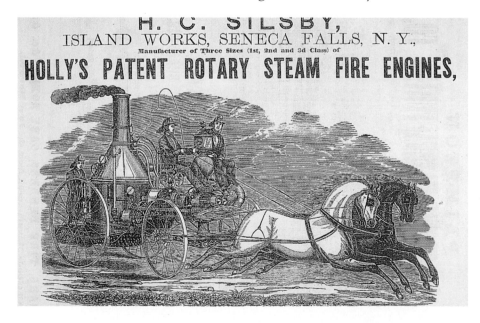

H. C. SILSBY,
ISLAND WORKS, SENECA FALLS, N. Y.,
Manufacturer of Three Sizes (1st, 2nd and 3d Class) of
HOLLY'S PATENT ROTARY STEAM FIRE ENGINES,

hand pumper that D.L. Farnham of Cincinnati had built. He coupled its horizontal double-acting piston pump and air chamber to his improved Buchanan boiler with a sheet-iron furnace. For the steam cylinders to power the pump, and a steam-propelled chassis, he turned to the brothers Alexander, Edmundson, and Finley Latta, whose Buckeye Works at 179 Race Street built everything from machine tools and stationary steam engines to complete railroad locomotives.

"Shawk built a massive wooden frame around the improved Buchanan boiler, Farnham pump, and Latta steam engine. Then (Alexander) Latta, Shawk, (councilman Miles) Greenwood, and (Cincinnati volunteer fire chief R.G.) Bray mounted the entire contraption on the carriage of one of the city's retired hand-drawn ladder wagons. Assembly of all of these components was performed at Miles Greenwood's Eagle Iron Works, 396 Walnut Street, during the winter of 1851-1852.

". . . On March 2, 1852 . . . Abel Shawk demonstrated his prototype steam fire engine to the City Council and about 3,000 citizens."

The test was a success, and after much pressure by city councilman Joe Ross, the city council appropriated $5,000 to Abel Shawk and Alexander Latta to build a steam fire engine for the city. Cincinnati's fire chief, R.G. Bray, donated another $5,000 from his own pocket.

The engine was built in the factory of Mr. John McGowan on Ohio Avenue. Completed December 15, 1852, the world's first successful steam fire engine, dubbed the "Uncle Joe Ross," had its trail on January 1, 1853. The world's first fully paid, professional fire department was organized at Cincinnati on April 1, 1853, to run the steamer. Councilmen Miles Greenwood, a former volunteer fireman, was appointed the first fire chief of this new fire department.

So 1853 marks the most crucial turning point in the history of America's fire service, from volunteers operating fire engines using muscle power, to professionals operating fire engines using mechanical power. Although fire engines changed from steam to gasoline to diesel power, and firefighters now undergo much more extensive training, the modern era of firefighting really began with the innovations that the 1851 Cincinnati fire and riot spawned.

Steam, Horses, and Electricity

Now let's fast-forward yet another half century. The date is April 18, 1906. The place is San Francisco, CA.

Probably the most written-about generation in American history is slowing dying off. This generation, now aged 60 to 70, fought in the American Civil War of 1861-1865. They then headed to the Midwest in the 1870s to become farmers or cattle ranchers. They moved again in the 1880s, joining wagon trains bound for California and Oregon, many fighting Indians or hunting buffalo along the way.

Many of the Civil War/cowboy/Indian fighter generation made their fortunes building and operating railroads, manufacturing all sorts of products for world markets, or exporting American-made agricultural and manufactured goods and importing from Europe and Asia. In California, some owned the steamship lines that imported and exported goods between America and Asia.

By the turn of the 20th century, this prosperity had built San Francisco into one of America's fastest growing and most densely populated cities. And one with a high risk of fire. Indeed, San Francisco had burned down and been rebuilt six times between December, 1849, and June, 1851.

By 1906, the technology of the fire service had changed greatly since Cincinnati introduced the paid fire department and the steam fire engine half a century earlier. Paid fire departments now operated in two, or even three shifts, called platoons.

Steam fire engines had abandoned the heavy, clunky steam propulsion mechanism in favor of two or three well-trained fire horses, and the steam boiler now operated only the pump. Boilers had gotten faster, generating steam from cold water in 3 minutes, one-third the time of the first steam fire engine boilers. Of course, steamers were kept coupled to stationary heaters in the firehouse, so that the boiler was hot at all times, and rarely had to start from cold water, so they had steam in even less than the three

Cast-iron Gamewell fire alarm box, City of San Jose #4238, hung on the exterior wall of a downtown brewery from 1916 until sometime in the 1940s, when the brewery closed its doors. The city let the brewery owner keep the fire alarm box, and he later gave it to his son. That same son, then in his 80s, sold the 75lb box around 1993 to a local dealer in antiques, who in turn sold it to the author in 1995. This box, with jewel-like brass clockwork inside and glass-encased key guard on the front, is quite a contrast from the less elegant, but 50lb-lighter, 1931 Gamewell box, also in the author's collection.

minutes required from a cold start. When a steamer rolled out of the fire station, it automatically uncoupled from the connections to the stationary heater.

After the American Civil War, the iron locomotive tire that summoned firemen went the way of the earlier watchmen's rattle, with the coming of a new technology: the electric fire-alarm telegraph. In the event of a fire, a citizen pulled down a hook on a fire alarm box at the nearest street corner. A gear inside the alarm-box turned, and as it did, a metal tab struck each gear tooth. Each box with a city had a unique pattern of teeth.

For example, the author owns a 1931 fire-alarm box from Fresno, CA, with the number 5327, which means that the gear has five teeth, then a gap, then three teeth, another gap, two teeth, another gap, seven teeth, and a final gap. As the gear turned, a telegraph signal sent five electric pulses, then paused, then sent three pulses, pause, then two, pause, and finally seven pulses. This clicking of a tab against gear teeth is remarkably similar to the primitive technology of the watchman's rattle, except that the gear motion results in an electrical signal rather than a clatter of wood on wood. However, if you open a fire alarm box while it is transmitting, and listen to the gear turning, the clatter is very similar to that of the watchman's rattle.

The central fire alarm telegraph office (in downtown Fresno in the case of my alarm box) received these signals from every alarm box in the city. The telegraph operator had printed cards telling him the street location of that particular box, and which engine and truck companies to dispatch to that location. He would then retransmit that signal to the appropriate fire station or stations. There, a gong on the firehouse wall would receive the retransmitted signal, and ring out the number: in this case, five gongs, pause, three gongs, pause, two gongs, pause, seven gongs, pause, then repeat the pattern two more times.

Firefighters upstairs, sleeping in the bunkroom, would often count the bells even before they fully woke-up. The more experienced ones might even know where that box was located. While the gongs were clanging and the firefighters were slipping into the boots, coats, and helmets kept always ready at the sides of their beds, a fire-

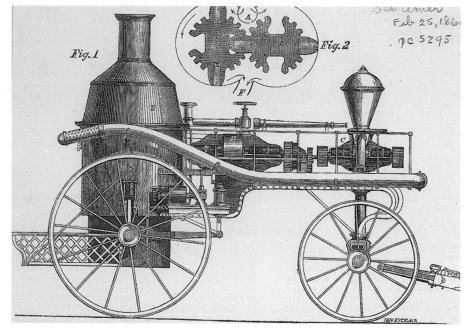

alarm tape register punched holes in a paper tape. First five holes, then a gap, three holes, another gap, two holes, another gap, and seven holes. Like the gong, the tape repeated the pattern twice more.

The watch-desk commander at the firehouse would look at the punched holes in the paper tape, and look up the printed card matching that number. He could then determine where the box was, and what other engine and truck companies would also arrive on the scene. The captain of that firehouse could then decide where best to park his apparatus on arrival, so as not to obstruct the other fire companies that would also be arriving on the scene.

Meanwhile, the firemen, now awake and fully dressed, wrapped their arms around a brass pole connecting the upstairs bunkroom to the downtown apparatus floor. Wrapping their coat-sleeved arms around the pole prevented friction burns to their skin as they slid down, and acted as a brake to slow the force of gravity pushing them down the pole. It was much quicker than running down a flight of stairs.

As if the firehouse did not suffer enough organized bedlam from the clanging gong, the flurry of activity around the watch desk, and firefighters sliding one after another in rapid succession down the pole, the alarm also set the fire horses into action.

As soon as the alarm arrived in the firehouse, the signal automatically opened the doors of the horse stalls. The best trained horses sensed the signal coming in even before the first tap of the gong on the wall.

Silsby, Mynderse, and Co.'s steam fire engine, as seen in *Scientific American*, February 25, 1860. *Archive Photos*

19

FIRE EQUIPMENT

Right: The Borough of Richmond (Staten Island) in New York City used this 1905 Waterous horse-drawn gasoline pumper, now in the New York City Fire Museum.

Below: When the central fire-alarm telegraph office in a city received a fire alarm, these paper-tape registers recorded the number of the alarm box as a series of punched holes. From the telegraph console in background, a trained fire alarm operator retransmitted that alarm-box number to one or more fire stations, near the location of the alarm box from which the signal originated.

When the stall door opened, the well trained horses knew to take their assigned places in front of the steamer. In a three-horse team, this meant that each horse knew whether to stand at left, right, or center in front of the engine. Harnesses, suspended from the ceiling, automatically dropped over each horse, and a firefighter snapped or buckled the harness securely around the horse. In big American cities of the late 19th and early 20th century, such as San Francisco in 1906, firefighters prided themselves in having the horses hitched and ready to dash into the street in less than a minute after the first tap of the alarm gong at the fire station.

As the engine headed out the door, the engineer riding on the steamer's rear platform shoveled a special kind of soft coal, called "cannel" coal, into the fire box at the bottom of the boiler. By the time the engine arrived at the fire, it had a full head of steam to start the pump turning.

The method of procuring the fire equipment of 1906, such as fire-alarm telegraph systems, was far more complex than in the days of the night watchman's rattle. The mayor and city council had to vote the funds from the city treasury to obtain and wire-up the system. Often, a municipal bond had to be issued to finance it, and the funds for issuing the bonds had to be approved by public vote of the community's citizens. Then the city had to buy the fire alarm boxes, usually from Gamewell of Newton, MA, which controlled over 90% of the American municipal fire-alarm telegraph market (although there were a few other American fire-alarm box manufacturers, such as Horni). They also had to buy the pedestals for mounting the boxes on each street corner, the gongs and tape registers for each fire station, and the main console and retransmitters for the central fire alarm office.

Once Gamewell or another manufacturer delivered all of this equipment, the city had to hire a contractor to build a structure to house the central fire alarm office, which usually required too much equipment to fit in any existing fire station. Then the city's electrical department had to string wires between all those fire alarm boxes, the central alarm office, and the individual fire stations. A crew of fire alarm operators had to be hired and trained to staff the central fire alarm office, and at least one firefighter at each fire station had to be trained as a watch-desk commander to read and interpret the fire-alarm telegraph signals.

Although this method sounds complicated and expensive, it worked so well that Gamewell's advertising slogan, "A Box A Block," literally came true in every village, town, and city in America, for 120 years. Only with the advent of the 9-1-1 emergency telephone system in the 1980s did fire alarm boxes finally fall into disfavor. As smaller cities and towns began to replace their fire-alarm telegraphs with 9-1-1 systems, the old familiar Gamewell box became a hot collector's item. I was lucky to get my 5327 Gamewell box in the early 1980s, just as collectors became interested in them, and you could still buy one in mint condition for $50. Now, 15 years later, if I were to buy that same box from a collector, I would probably part with $200.

Even today, not every city has replaced its fire alarm boxes, a technology nearly unchanged in 130 years. In larger cities (such as New York, Chicago, and San Francisco), the 9-1-1 emergency number receives an extraordinarily high number of police, ambulance, and fire calls. Were it not for the aging fire-alarm telegraph systems that those cities still use in parallel with the phones, the volume of calls would overwhelm the capabilities of local 9-1-1 operators to promptly respond.

Fire hydrants, too, had changed in half a century. Massive steam-powered pumping stations supplied pressure to the water in the

As an experiment, in 1931 Motorola in Illinois installed its first two-way emergency radio in a then-new 1930 Ahrens-Fox quadruple-combination pumper and ladder truck at nearby River Forest, IL. Unfortunately, the groundbreaking radio does not survive, but the fire engine that once carried it does. This Ahrens-Fox Model U-N-4, serial number 3392, is now displayed at the Hall of Flame, America's largest fire museum, in Phoenix, AZ.

city mains, and fire hydrants were installed on nearly every street corner.

New types of fire engines had also been developed in the previous half-century. The aerial ladder, raised and lowered by means of a massive hand-cranked spring, allowed firefighters to rescue citizens from the multi-story tenement houses that sprang up in major American cities of the late 19th century. Water towers, raised and lowered by hand crank, chemical reaction, or water pressure, delivered huge volumes of water at great pressure to the upper stories of tall office buildings. San Francisco in 1906 had three water towers: an 1890 Hale, and 1898 and 1901 machines built by San Francisco Fire Department's own Henry Gorter. There is no evidence that any of these three towers saw use at San Francisco's 1906 Earthquake and Fire.

The soda-and-acid chemical engine used a mixture of bicarbonate of soda and sulfuric acid to cause a chemical reaction inside a copper tank, which forced water through a rubber hose onto a fire.

Fire hose, too, had changed. Gone was the leather hose with copper or brass rivets. Hose now came in 50ft lengths, 2.5in diameter, made of rubber surrounded by two woven cotton jackets, fitted with threaded brass couplings at each end. This type of hose was in use until lighter plastic hoses with aluminum couplings appeared in the 1970s and 1980s. Each length of hose weighed 50 pounds. Special hose wagons, resembling horse-drawn versions of today's pickup trucks, carried 1,000ft (20 lengths) or more of this hose. Firefighters attached this hose to the discharge gates of the steamer, and stretched multi-segment hose lines to the fire, while the steamer and its operator stood a safe distance from the blaze. On arrival at a fire, the horses were unhitched from the steamer and the hose wagon, and led away to a safe distance from the fire.

A Dutch innovation in how fires were fought had also taken hold in America. Instead of standing outside a burning building and throwing water through an open window or door, firefighters actually entered the burning building with their hose lines, sought out the heart of the fire, and sprayed the water directly into the jaws of the snarling beast called fire. The combination of this new firefighting tactic, and stricter building codes that favored brick and stone over wood construction, meant that fewer buildings burned to the ground now than they had 50, or even 25, years earlier.

Even the firefighter's clothing had changed. In the 1850s, the volunteer firefighter wore heavy black pants, a thick red woolen shirt, and a helmet that looked more like a gentleman's top hat for an evening at the opera than it did like today's fire helmets. He also wore a bandana around his neck, which he could place over his nose and mouth to filter-out smoke pouring out of a burning building, or soak in water and wrap around his neck to prevent blisters.

By 1906, firefighters wore heavy black rubber rain coats called "turnout" coats, canvas or other durable trousers, rubber boots that they could pull all the way up to their knees, and leather fire helmets to protect their heads from objects that might fall from the burning ceilings. These helmets had downward-curved brims in the back, so that the water which other firefighters were spraying at the fire would run down the outside of the back of the firefighter's coat, instead of down his neck.

Such was the state of fire service technology when the ground under San Francisco began to quake at 5:12 on the morning of April 18, 1906. The city had suffered a major earthquake in 1868, and smaller ones in 1892 and 1898, but nothing that would have prepared its citizens for the events of the next few days.

Even a dire prediction by the National Board of Fire Underwriters (NBFU) just seven months earlier, in October, 1905, had not prepared the city for what was to come. This NBFU report stated that San Francisco's 36-million gallon per day water works could not supply the fire hydrants with enough water at sufficient pressure to guarantee the city's safety from fire. But the city's firefighters had always successfully battled any fire that came along, so local politicians were apathetic about spending the money to improve the hydrant system.

In the produce district, men were unloading fresh vegetables for sale in stores and outdoor markets. Their horses, with a sense of foreboding, shifted and whinnied. Seconds later, a deep rumbling starting in the distance and grew closer. Those looking up the street could see an undulating wave rippling through the street as the rumbling noise grew closer and louder. As the rippling wave passed, it quickly toppled even the sturdiest of brick structures, crushing their occupants, and burying those in the street beneath heaps of rubble. As buildings fell, they creaked and groaned, adding to the noise like rushing wind and crashing ocean waves that the roaring earthquake was making.

Then, as quickly as the rumbling shock wave hit, it was gone, and an eerie silence momentarily filled San Francisco's streets. Soon enough, however, the shrieks and screams of the quake's severely injured victims began to fill the silent void.

South of Market Street, row on row of homes, offices, stores, and flimsy wooden hotels lay in ruins. Those homes that were not completely ruined, were knocked off kilter just enough that doors and windows would not open to release their trapped occupants.

The chimney stacks of the California Hotel crashed through the roof of the San Francisco Fire Department headquarters next door. Fire chief Dennis Sullivan, a veteran of 26 years in the fire department, was struck by a piece of this chimney, leaving him with severe chest and head wounds.

As a result, over the next few days, the city would not benefit from Sullivan's knowledge and experience in its time of greatest need. Mortally injured, Chief Sullivan would die just three days later. Still, this handicap was somewhat overcome when Assistant Chief Daugherty, nearly as experienced as Sullivan, took command.

As a few of the city's braver souls began to wander back into their homes, a less severe but still extremely strong second shock wave rumbled through the city at 5:26am. Over the next few hours, several smaller after-shocks followed, but the first two big waves caused most of the early-morning deaths and destruction.

The earthquake's path had extended 20 to 40 miles west from the Pacific Ocean, and 200 miles north to south, from Salinas to Fort Bragg. For most of these communities, death and destruction, although terrible, had been swift. But for San Francisco, the greatest death and destruction still lay ahead.

Even before the initial jolt subsided, no fewer than 50 separate fires started in San Francisco, many from dislodged natural-gas lines, broken street lights, and downed overhead electric wires, such as those used for telephones, trolleys, and even fire-alarm telegraph systems. More fires probably started in various homes, but were quickly extinguished by the homeowners before causing any serious damage.

By noon on April 18, San Francisco firefighters had extinguished most of the 50 fires to which they responded that morning.

As if disabling their fire chief was not enough, the quake had delivered the city's firefighters another handicap: the fire alarm headquarters, on Brenham Place in Chinatown, was reduced to rubble. No fire station in the city could receive the alarm bells when a citizen pulled the hook on a street-corner alarm box. Even this handicap was overcome somewhat. Instead of waiting to be summoned to a fire, firefighters patrolled their neighborhoods with their equipment, looking for fires to extinguish.

By 6:00am, the city's firefighters learned of their most severe handicap, one that they could not overcome. As they moved from fire to fire, opening valves on fire hydrants, only trickles of water came out. The quake had damaged the city's water mains, leaving the hydrants dry. In most cases, either enough water remained in the hydrant to extinguish the fire, or another water source was found. Four separate fires north of Market Street, between Sansome and Washington, were contained this way. But south of Market, a dozen separate fires were spreading through row on row of ancient wooden structures, with not a drop of water available to stop the relentless advance of the flames.

By noon, the remaining fires in the South of Market district, and in several other sections of the city, were out of control.

Resourceful firefighters found many creative ways to combat the fires. Some dug up the streets to use residual water in the mains. Others used water in the city sewers. Along the waterfront, firefighters battled blazes using salt water from the bay. Some recalled that the city still had 23 ancient cisterns, like giant wells, holding from 16,000 gallons to 100,000 gallons each.

Some quick-thinking citizens had filled buckets and bathtubs with water just after the first shock, and those citizens were able to save their houses when all the buildings around them burned to the ground. Some even opened casks of wine and vinegar to at least have some sort of liquid to protect their homes from the advancing flames.

General Funston, in command of the Army at San Francisco, directed the dynamiting of buildings in the path of the flames. This action needlessly destroyed some buildings that the flames never reached. In most cases, it had little or no effect on the fires, as they either consumed the rubble as fuel, jumped over the rubble to burn the next building in line, or changed direction to burn a building in a different path.

By noon, most of the South of Market area had burned, and several other neighborhoods were severely damaged. Again, the worst destruction was yet to come.

About 9:00am, a woman in Hayes Valley, the neighborhood bounded by Van Ness, Octavia, McAllister, and Market streets, decided to cook breakfast. Her house was on Hayes, only a few doors west of Gough. She was unaware that the earthquake had damaged her flue, and sparks from the fire that she started in the kitchen stove, soon set her walls ablaze.

All of the city's fire companies were busy fighting fires in other neighborhoods, so no help was available to check the advancing flames of what was aptly dubbed the "Ham and Eggs Fire." By 1:00pm, this blaze had destroyed a hotel, a church, and even a college. It would eventually spread to consume City Hall, too, before a heroic effort by firefighters stopped its advance along Octavia Street and Golden Gate Avenue before dawn on the 19th. The main South of Market fire was losing momentum, too, having consumed all the fuel in its path.

The final threat was on the 20th, when it looked as if the main fire might spread to the wharves. But the fire did not shift, and firefighters had doused all the waterfront buildings that were endangered. A rain that evening extinguished the remaining patches of fire, and by the morning of the 21st, there was no more fire, only ruins. 490 city blocks, or 2,831 acres, lay in ashes. 500 lives were lost, and 28,000 buildings destroyed, for a loss of $350 million. Compare that to the Great Chicago Fire of October 9-11, 1871: 250 dead, 17,000 buildings destroyed, and an $88 million loss.

Gasoline Motors and Radios

The 20th century has seen not only an explosion of new technologies, but also acceleration in the rate at which technology advances. The same applies to fire service technology.

While in previous sections, we saw fire service technology change but little in 50 or even 100 years, we need to advance our imaginary video only 25 years to see a great change in the fire service from the picture in 1906.

The setting this time is River Forest, IL, a suburb of Chicago, in the year 1931. The horse stalls, hanging harnesses, and haylofts have disappeared from the fire station.

The massive 75lb cast-iron street corner fire alarm boxes have only recently given way to aluminum fire alarm boxes weighing just 25lb each. These new boxes had non-interrupting circuits, so that if power was out or a signal from another alarm box was being transmitted, the signal was held until it could be successfully received at headquarters. The author has turned off power to his 1931 Gamewell fire alarm box, Fresno #5327, in mid-signal, only to have it continue transmitting precisely where it left off when I turned power back on again even hours later. With these new boxes, the fire alarm signal would always get through, no matter what.

The need for one fire engine to carry a pump, another to carry hose, and a third to carry ladders, has also vanished. So, too, has the soda-and-acid chemical engine, using a dangerous mixture of bicarbonate of soda and sulfuric acid to cause a chemical reaction that forced water through a hose onto a fire. Just months earlier, River Forest had purchased a brand-new Ahrens-Fox "Quad," or quadruple combination. It carried a 1,000-gallon per minute piston pump, 1,000ft of 2.5in hose, and a 60-gallon water tank, or "booster" tank that supplied water through a one-inch rubber hose. It also carried the following ladders: one 55ft, two 35ft, and one each 28ft, 24ft, 20ft, 18ft, 16ft, 12ft, and 10ft.

The truck has button-tufted leather seats, but no windshield, doors, or roof. Firefighters joke that it has "Armstrong" steering, because you need strong arms to turn the 30-plus feet long rig around corners. It has two-wheel mechanical brakes, assisted by a vacuum booster brake, so you have to plan your stops about a block ahead. It has no siren, but it has a 12in locomotive bell at top center of the wooden dashboard, and a shrill whistle under the

When American LaFrance of Elmira, NY, built one of its popular 700-series pumpers for Merced, CA, in 1949, two-way radios were still a rarity in the fire service. Today, in the hands of collector Leonard Williams of Sunnyvale, CA, the pumper's aging Motorola radio is still in good working order.

chassis, powered by gases from the exhaust pipe.

The power for this four-in-one fire engine is also a big change from 1906. It uses Ahrens-Fox's own six-cylinder, T-head gasoline engine, with massive 5.5in pistons and a 7in stroke, for a total piston displacement of 998 cubic inches, and a brake horsepower of 110. The engine exhaust is open cutout style, no muffler, so the noise is nearly deafening for the driver and the Captain beside him. A 40 gallon gasoline tank, above and behind the driver, keeps the engine fueled.

Arriving at a fire, the driver switches the engine into pump gear. This disengages power to the rear axle, and uses the same gasoline motor, now turning at much lower RPMs, to power the pump instead.

Not far from River Forest, engineers at Motorola have just developed a two-way radio that can run off a vehicle battery. As an experiment, they install it in the new Ahrens-Fox quad at River Forest. Firefighters en route to a fire, or at the fire scene, can now instantly receive instructions from fire headquarters, and transmit information about fire-scene conditions back to headquarters. After the trial, Motorola engineers remove the radio, and bring it back to Motorola. The trial has taught them much about what they did right, but also what bugs they still need to work out of the system.

A few other fire departments installed two-way radios in their fire engines in the late 1930s, and Ahrens-Fox even did a factory installation of a two-way radio in one of their pumpers in 1939. But two-way radios in fire engines, and the portable hand-held "walkie-talkie" radios that today's firefighters carry with them everywhere, did not gain popularity until the 1960s.

Diesels, 9-1-1, and Pollution Controls

Let's fast-forward our video of glimpses of American fire service history one last time. It is now 1996 in San Jose, California. A trained 9-1-1 emergency operator sits in front of a console containing a telephone and a computer monitor. This operator's telephone rings, and a panic-stricken child shrieks, "Our house is on fire!" and hangs up.

Instantly, the computer monitor shows the 9-1-1 operator that the call originated from 1121 Loupe Avenue.

She punches a redial button on her console. This rings the phone from which the call originated. A woman answers.

"Ma'am," the operator says, "this is the 9-1-1 Emergency operator. Your child just called us to report that your house is on fire. Is your house on fire, ma'am?"

"Yes, it is."

"Does anyone there need medical attention?"

"No, my son and I are OK. But our kitchen is on fire, and I'm afraid it's going to spread to our living room."

"Can you and your son get out of the house quickly?"

"We're already outside. We're talking to you on a cellular phone."

"Ma'am, is anybody still inside the house?"

"Nobody. But all my belongings are inside."

"Ma'am, please stay outside. Do not go back into your house to retrieve your belongings. Okay?"

"Okay."

The 9-1-1 operator now pushes another button on her console, to contact the San Jose Fire Department Dispatcher.

"This is 9-1-1 Emergency. We have a structure fire in the kitchen at Eleven-Twenty-One Loupe Avenue. Occupants are outside, and do not require medical attention."

The San Jose Fire Department dispatcher enters the address, 1121 Loupe Avenue, into her computer console. The computer tells her map coordinates for that address, and also tells her that the nearest available fire engine not already out on another call is Engine 26, on Tully Road near the Santa Clara County Fairgrounds.

Inside Station 26, a shrill tone warns the firefighters of an incoming call. "San Jose Fire, Engine 26, structure fire, in the kitchen, Eleven-Twenty-One Loupe Avenue. One-One-Two-One Loupe Avenue. Cross street is McLaughlin. Thomas Brothers map, page sixty-eight, A One. Engine 26, you are first due."

The watch-desk commander at Station 26 punches the map coordinates, 68-A-1, into his computer, and instantly prints out a map and directions.

Some of the firefighters are in the kitchen preparing dinner. Others are watching a ball game on TV. Still others are in the rec room, lifting weights. For the past twenty years, a few of the firefighters in this male-dominated firefighting fraternity have been women, and one of them is on duty at Station 26 on this shift.

They all head for the shiny red Hi-Tech pumper, delivered earlier that year. They grab their turnout coats, which are treated with flame-resistant Nomex chemical, and their nearly puncture-proof helmets, which hang from hooks on the wall. They put on their coats and helmets as they walk. The Captain grabs the map and directions from the watch-desk commander before climbing into the right front cab seat. He will guide the driver to the fire.

The pumper has a 2,000-gallon per minute centrifugal pump. Its motive power is a Detroit diesel, coupled to an Allison automatic transmission. The truck has power steering and air brakes. Inside the enclosed and air-conditioned five-man cab, the seats are plushly cushioned, and equipped with shoulder and waist seat belts.

As each firefighter climbs into his or her assigned seat in the cab, he snaps his turnout coat closed, slips into the bunker pants and rubber boots in front of his seat, and pulls protective gloves onto his hands. A filled air bottle hangs behind his seat, and

After the 9-1-1 call, the operator calls the fire department dispatcher, who sends the nearest fire engine. As the engine approaches the fire, the task is assessed and then acted upon: and the dangerous part begins. *Image Bank*

The strain, physical exertion, and heat saps even the most experienced. *Image Bank*

he straps it onto his back, then places the clear plastic mask over his face. The mask connects to the air tank via a corrugated rubber hose. Next, he straps on head-phones, which not only protect his hearing from the roaring diesel engine and electronic warning devices, but also let him listen-in to radio communications between the driver, the Captain, and the fire alarm dispatcher.

Finally, he pulls his portable walkie-talkie out of the pocket of his turnout coat and switches it on, so it will be ready when he arrives at the fire. He reaches in another pocket and switches a flashlight on and off, to be sure it is working. By the time he arrives at the fire, he is ready to jump out and get to work.

As the engine rolls out the door, a large yellow plastic tube, suspended from the ceiling, snaps off of the exhaust pipe. This tube prevents diesel fumes from filling the fire station, where firefighters might breathe the toxic smoke. Instead, the diesel exhaust travels through the tube and is safely vented to the outdoors.

As the big Hi-Tech rolls into the street, the driver switches on the flashing red lights on the cab roof, and an array of electronic sirens that wail, yelp, and warble. The Captain pulls a chain hanging from the cab roof, activating a loud two-tone air horn.

As Engine 26 speeds to the fire, the San Jose Fire dispatcher radios additional information, for which there was no time during the original dispatch.

"San Jose Fire, Engine 26, R.P. (reporting person) reports that she and her son are outside. Nobody is in the structure. No medical attention required."

As the engine approaches the fire, the Captain and the firefighters size-up the scene, and determine what each must do on arrival. They each remove their headphones, toss them on their seats, and jump out of the cab. One firefighter swings out the free end of the large, plastic-covered soft-suction hose (the other end is already attached to the pump), couples it to the hydrant on the corner, and opens the valve on top of the hydrant, using a special hexagonal wrench. Two firefighters pull a 1.5in hose line from a compartment atop the pump, attach it to a pump outlet, and head into the burning house. One of the two firefighters in the

kitchen shouts "water!" into his walkie-talkie, and the female firefighter, operating the pump, opens the discharge gate. The fire hose swells as it fills with water.

A few minutes later, the two firefighters emerge, the fire extinguished, and damage held to a few black marks on one kitchen wall. The two firefighters take a large fan out of a compartment of their engine, and place it in the open front door to exhaust all remaining smoke from the house. After a few minutes, they shut off the fan and return it to its compartment.

The pump operator closes the discharge gate valve and opens a drain under the pump. She then disconnects the 1.5in hose from the pump.

The two firefighters who just extinguished the fire now uncouple all the hoses they just used. Each places one end of a length of hose over his shoulder, and walks its length, using gravity to drain all water from inside the hose. Both firefighters then roll the used hoses into doughnut-shaped rolls, and stack them on top of the pumper.

On the way home, all of the firefighters remove their bunker pants and boots, and place them in their proper location in front of the seat.

Back in quarters, the two firefighters who were inside the burning house, put the used hose rolls into a big drier that bakes them until dry, before recoupling the hoses and returning them to their proper compartments on the pumper.

The solitary female firefighter takes the used air bottles to a back room, where she refills them from an air tank. She then stows them back in their hangers behind each cab seat. Meanwhile, the watch-desk commander snaps the yellow diesel-fume hose back over the end of the exhaust pipe, to vent the smoke the next time the big diesel engine roars to life.

After everyone hangs their turnout coats and helmets back in their proper places, dinner preparations resume where they left off. A few gather around the TV, which was never turned off, to continue watching the ball game.

This concludes our brief video tour of American fire service history. The chapters that follow will examine, in greater detail, the fire equipment used in each of the eras we just visited.

CHAPTER ONE

Fire Equipment in the Colonial Era 1608-1790

Hero of Alexandria invented the first known fire engine in ancient Roman times. It consisted of two pistons that forced water out on their down strokes, and an air chamber to regulate the flow of water between piston strokes. In 1548, Hero's writings, including a description of this fire engine, were rediscovered, translated into German, and published.

The timing of this publication could not have been better. Various European countries, chiefly Holland and England, were just beginning to colonize a newly discovered continent called America, and the settlers and their new villages would soon enough need the protection of a fire engine.

The first recorded fire in American history occurred in colonial Jamestown, VA, in February of 1608. 100 Europeans founded the colony the previous May; but that first winter, 62 died of disease. Most of the remaining 38 colonists were ill or disabled when eight more colonists arrived in January, 1608, bringing with them fresh supplies and renewed optimism for the colony's future.

That optimism lasted less than a month, when the community blockhouse, built of rough-hewn lumber from the area's then-plentiful forests, caught fire. The blaze spread to burn not only most of the settlers crude thatched-roof homes, but as Captain John Smith reported, "most of our apparel, lodging, and private provision." Left homeless, many froze to death, and those who

survived nearly starved the following winter. In 1609, local Indians, determined to rid themselves of the English invaders, began killing Jamestown settlers one by one in the nearby woods. In 1610, the few remaining settlers temporarily abandoned Jamestown and returned to England.

Fifteen years later and many miles north, the colonists who had founded Plymouth, MA, in 1620, suffered their own devastating fire in November, 1623. Some sailors had lit a fire in their chimney when it spread to the thatched roof, and from there burned down 3 more houses, and only with great effort was the community's food storehouse saved from destruction.

Some of the Plymouth Puritans created a new settlement, Boston, in 1630. Eight months after their arrival, the few original homes in Boston burned down in a repeat of the 1623 fire. Governor Winthrop of the Massachusetts colony wrote of this 1631 fire: "About noon the chimney of Mr. Thomas Sharp's house in Boston took fire, the splinters being not clayed at the top, and taking the thatch, burnt it down. The wind being Northwest, drove the fire to Mr. Coulbourn's house, and burnt that down also."

As a result, in 1631, the community leaders of Boston passed the nation's first fire-safety building code, outlawing wooden chimneys and thatched roofs, and favoring instead brick and stone construction.

Before a single Puritan ever set foot in

New England, however, some Dutch sailors under the command of Captain Adriaen Block, who were profitably trading with Indians of the Hudson River region, lost a ship, the Tiger, in a 1613 fire at the mouth of the Hudson River off Manhattan Island. The sailors, having no way to return home until another Dutch ship could arrive, founded the community of New Amsterdam, which eventually became today's New York City. Their crude houses, without floors, stood where Broadway and Exchange Place now merge. These homes, like most in earliest Colonial America, consisted of stone or reed-lined wood chimneys, frameworks of bent-sapling boughs, and thatched (dried grass and straw) roofs.

The Manhattan tribe of Indians brought these sailor/settlers food, and helped them hew the wood beams to make a new ship. Returning to Holland, their tales of Manhattan Island encouraged more Dutch settlers to come over. In 1615, these Dutch traders built America's first commercial warehouse, the United Netherlands Company, on lower Manhattan, importing goods from Holland, and exporting furs and foods brought by the Indians. In 1628, one of New Amsterdam's homes burned to the ground when an overheated wood chimney set fire to the thatched roof.

In 1647, New Amsterdam's Dutch governor, Peter Stuyvesant, passed an ordinance similar to the one Boston enacted in 1631, prohibiting wood chimneys and thatched roofs. The following year, four fire wardens were appointed, to inspect all buildings in the community and to fine any violators of the fire codes. This quartet of fire wardens were America's first organized fire department, and came from diverse backgrounds, showing that even 300 years ago, America was already on its way to becoming the "Great Melting Pot." Adrian Keyser was an officer of the Dutch West India Company. Martin Krigier owned a tavern. Thomas Hall had been a wealthy Englishman before becoming a war prisoner of the Dutch. George Wolsey was an English trader who came over on the Mayflower and, unhappy with the repressive Plymouth colony, threw in his lot with the Dutch on Manhattan Island. The job of these four was not fire-fighting, but fire prevention, and collecting the three guilder fine for building code violations, and the 25 guilder fine from anyone whose negligence caused a fire. The New Amsterdam government used these fines to purchase buckets, ladders, and hooks for pulling down burning roofs.

These fires of the earliest colonial era had in common the fact that there was not one fire engine, and few buckets, in all of America. Even where buckets were available, the intense heat of the burning grass and straw roofs kept would-be firefighters from getting close enough to toss the water onto the fire.

In each of these early colonies, after losing nearly everything to fire, new ordinances required every citizen to aid the community whenever another fire struck.

The colonial citizen-firefighter's main weapon against fire was the fire bucket that he was required to keep at the ready at all times, on the outside of his home. Made of leather, with hand-sewn seams, each bucket held 2.5 to 3 gallons of water, and weighed 25 to 30lb when full (a gallon of water weighs 8.3lb). The bucket handle was made of hemp rope, encased in leather, and attached to the top of the bucket's side walls with brass fasteners. About 1820, copper or brass rivets replaced hand stitching as a means of closing the bucket seams.

While leather buckets were more durable than wood, they were just as susceptible to rot. With the discovery of the rubber vulcanizing process in the 1840s, many volunteer fire brigades switched from leather to rubber buckets. Starting in the late 1700s, some commercial establishments used metal buckets, with rounded or pointed (conical) bottoms. Not only did the odd-shaped bottoms permit their owners to fill them with a larger volume of water or sand,

Colonial watchmen carried wooden rattles like this one, to summon citizens and their buckets when a building caught fire. *Ed Hass collection*

31

but the fact that they could not lay flat on a floor discouraged their use as mop buckets.

Organized volunteer fire brigades were virtually non-existent among early colonists. Instead, every citizen had his own bucket, with his name hand-lettered on it in oil paint. Some buckets also had the family coat of arms, a portrait of their owners, or other highly decorative art (such as an eagle) painted on them.

Bucket brigades passed filled buckets from a river, lake, horse-trough, or cistern (a hole in the ground, edged stone or brick, used like an oversized well or tank to catch and store rainwater). A second line of citizen-firefighters returned the empty buckets down a second line for refilling.

Today, original leather fire buckets of the colonial era are highly prized among collectors of American fire service memorabilia, particularly if they come with documentation that authenticates their original owners. As a rule, the more elaborate the painted decoration on the bucket, the more valuable it is in the collector market.

Items that detract from a genuine Colonial bucket's value include repaired or replaced handle, or cracks in the paint. The original paint is obviously more valued, but a bucket design that has been carefully retouched by a skilled craftsman is nearly as prized. If the bucket was originally painted by a famous artist, or bears a portrait of a famous person, it is generally more prized by collectors. An 1806 fire bucket, sporting an elaborate painting of the messenger-god Mercury blowing a trumpet, fetched $30,000 at an auction in the 1980s. However, you can still find the more mundane buckets that bear only the owner's name, in fair to good condition, for $100 to $500.

Beware of fakes, however: in the 1950s, famed fire helmet maker Cairns offered a replica leather fire bucket, and there are those who will rough-up one of these reproductions and try to pass them off as an original. If the leather or paint looks too good to be 200-300 years old, or the paint is not oil-based, it is not the genuine item. If there is no accompanying documentation, or the documents could not have been produced with the quill pens and crude printing presses of the era, be wary.

Another highly prized early American firefighter's tool is the watchman's wood rattle. I have two in my collection, but both were made long after the colonial era. One had two reeds and a round handle, but one of the reeds long since broke off. I found this one in a Midwestern antiques store. It had no date or accompanying documentation, but it is of a style popular in the late 18th and early 19th century. However, it does not appear to have been hand-carved by the watchman himself, as there are no signs of tool marks. It might be a reproduction, or if it is genuine, it was probably made in one of the era's small factories that employed a dozen or more wood carvers to make inexpensive wood products. The round, tapered handle may have been made on one of the first crude lathes that such establishments used to make table and chair legs. It has no varnish or other protective coating. Were it authentic and in mint condition, it would be valued at about $85 to $100.

My other watchman's rattle has only a single reed, and an octagonal handle. It is a much simpler design, and bears hand-tooling marks. The gear and reed are attached to the framework and handle using simple wood pegs, or dowel pins. The handle bears a hand-carved date of 1813, and the varnish is very dark and worn from age. The California antiques dealer from whom I bought it showed me, but would not part with, documentation from the 1890s showing that the then-owner obtained it from his grandfather. From this document, the rattle's condition, and the simple techniques used in its construction, I feel pretty confident that it really was made in 1813, decades before sledge-hammered iron locomotive tires, fire-alarm telegraphs, and telephones made the watchman's fire rattle obsolete. Sadly, some unthinking owner in the 1920s, evidently fearing that the 110-year-old wooden dowel pins were insufficient to hold the framework together any longer, inserted modern screws and nuts. But those pins are still there, proving the alteration unnecessary. This alteration detracts from the $125 to $150 value it might otherwise have on today's collector market. Of course, this added hardware is now 70 years old itself, so it is too antique to remove at this late date.

Because my two watchman's fire rattles are not in mint original condition, they probably would not bring much on the collector's market. But to me, the fact that they were used, abused, and altered makes their histories even more colorful and interesting.

Watchmen's rattles were usually made of mahogany, oak, or maple. They consisted of a carved, or in some cases lathe-turned, handle at one end; one or two wooden reeds; a wooden cog (gear) that the reeds struck as the rattle spun; and a supporting wooden framework. Some had brass plates at the end opposite the handle, the added weight providing momentum to keep the rattle spinning. The versions with these plates bring a slightly higher resale price than those that, like the two I have, lack such adornment.

If you plan to add a watchman's rattle to your collection, first pick it up and spin it. It should clack loudly, with no cog teeth missing or obvious cracks in the cog, reed, or supporting framework. There should be no evidence, such as obvious tool marks, that the rattle was disassembled to replace a reed or cog. The wood should be a hardwood, and the entire assembly held together with wooden pins or obviously hand-made (not machine-made) screws. If bolts and nuts were used instead of wood screws, they should be of the old square-head variety (hexagonal-head bolts and nuts are a very recent invention). The condition of the wood and its finish should be uniform; no part should look newer or less used than any other part. And as with leather buckets, authenticating documents add to the rattle's value.

For nearly 10 years, I owned a pair of hand-drawn hose reel carts that I would show at a variety of historic events all over California. I would also bring along my replica of a colonial leather fire bucket, and my evidently genuine 1813 rattle. I would then proceed to demonstrate how colonial firefighters were summoned, and what tools and methods they used to combat a blaze. Whether you collect modern replicas, or the genuine rattles and buckets, putting on such public demonstrations can be a lot of fun for you, and educational for the children and adults who watch you.

Another tool in the Colonial firefighter's arsenal was a simple swab mop. The thick mop could be used to smother a fire and, when wet, could put more water on the fire.

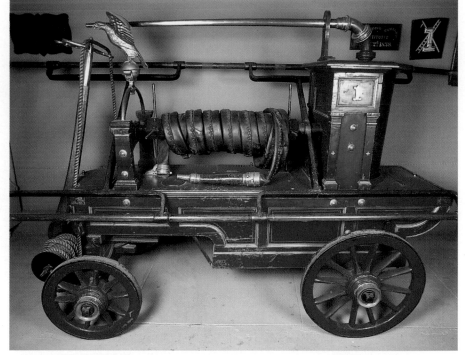

A rope tied to one end of a simple metal hook, when thrown, could quickly tear down a burning thatched roof before flying embers could spread to other houses. Once on the ground, it was easier to drown the burning straw and grass with buckets of water. Today's highly trained professional firefighters disdainfully refer to this tactic as "surround and drown," but this was the only firefighting method available in the early 1600s.

Fires continued to be a problem in early America's growing cities. In 1645, Boston suffered a gunpowder explosion that burned several nearby buildings and shook much of the city. Incredibly, city officials did not react to this disaster by requiring homeowners to have buckets, ladders, hooks, and axes. The city had no watchmen to look out for fires, and no bells to summon the aid of citizens in event of fire. The English had developed a giant syringe which could be filled with water drawn from a well, and then squirted on a fire; a few of these were even mounted on two wheels and provided with a hand crank to control their motion. But no such engines had yet been imported to Boston or any other English colony across the Atlantic.

And so the stage was set for the Great Boston Fire of 1653.

By this time, Boston was already the biggest city in the Massachusetts Bay colony, bigger than Plymouth or Salem. Saw mills in New Hampshire and Maine, and

Typical of hand pumpers of the Revolutionary War era is Engine 1, now at New York City Fire Museum.

brick kilns in nearby Dorchester, provided the materials for the many new homes and commercial establishments that were constantly being built. Houses made of tree boughs, with wood chimneys and thatched roofs, were rapidly giving way to clapboard houses with brick chimneys, shingle roofs, and plaster interior wall coverings. A few houses were even starting to feature glass windows, but these windows were small, due to the high price of glass, the stiff English tax on such materials, and the bitter cold that seemed to enter homes with more glass surface.

On a cold night in January, 1653, James Everill awakened Boston resident with knocks on their doors and shouts of "Fire! Fire!" A bright glow of a fire already involved several docks and wooden warehouses at the foot of King Street. Everill had spotted the fire because he owned a nearby brewhouse, warehouse, and wharf, all of which were in the path of the advancing fire. Boston residents David Sellick and Sam Cole were the first to join Everill on the scene, but they only had time to save a very few of Everill's belongings before the fire engulfed Everill's house and marched relentlessly onward. From there, the fire soon burned Hudson's tavern and wharf, the Oliver and Pierce family homes, a barn, and a dozen more buildings along King Street. Soon, the whole town turned out, watching helplessly as the fire spread. The town selectmen (similar to a modern city council) quickly organized citizens into small groups to remove food, clothing, livestock, ammunition, utensils, and furniture from homes in the path of the advancing conflagration, and move them to nearby open meadows. A few citizens managed to fill buckets of water from wells and wet down houses in the path of a fire. A few more brought ladders to salvage possessions from second floors of houses and commercial buildings. As the thick, black smoke rolling over the top of the city dropped flaming embers on houses below, blankets were soaked with water and tossed onto rooftops to save as many houses as possible.

Massachusetts Colony's Governor Endicott and the town selectmen directed citizens to tear down houses on New Street, and either cart away or soak the debris from each house, creating a wide enough fire break to stop the fire's advance. For the older homes with thatched roofs, yanking on two ropes with hooks on their ends quickly demolished all but the chimneys. The newer, massive clapboard houses took considerably more effort to tear down. Meanwhile, the work of salvaging property from houses stopped, because the heat was too intense, even four houses away from the advancing flames.

The fire took a turn for the worse when it reached a barrel of gunpowder stored at Sam Cole's "house of Publick intertainment." As citizens ran from that explosion, the fire reached another barrel of gunpowder in a private home on King Street. All citizens were ordered out of the immediate fire area, as it was too dangerous.

The firebreak on New Street proved ineffective, as the fire jumped over the torn-down homes to ignite the roof of Reverend John Wilson's house. Naturally he, too, stored gunpowder, starting a second major fire that spread to several nearby buildings and raced toward the first fire as the original blaze raced toward the new one. Three young children in the Sheath family somehow slept through the fire, and nobody had noticed they were missing until they, and their house, were consumed in flames.

Citizens armed with buckets of water and swab mops stood between the two fires as they moved toward each other, and managed to drown and beat-out both fires before they could join forces and move in a new direction.

The next day, an emergency meeting of the town council decreed that every Boston citizen was to have a ladder that could reach to the top ridge of his roof, and a long pole with a swab at the end to quench a roof fire. Further, four chains with strong iron hooks on their ends were to be kept at the meeting house (city hall) to pull down burning roofs. A fire warden, carrying a hand bell, was to patrol the streets between 10:00 PM and 5:00 AM, to warn citizens in event of fire.

Other Colonial American cities had already passed similar ordinances. What made the Great Boston fire unique was that, within a year afterward, the city ordered the first fire engine ever used in America. The 1654 Boston fire engine was built in England by Joseph Jynks, and brought to

Boston by ship. It is not clear which of the two types of engines then popular in England was sent to Boston. It may have been a portable reservoir, or "squirt," mounted on two wheels. More likely, it was an awkward and heavy iron cylindrical syringe, three feet long, which three men carried to a fire. At the fire, two men pulled back the plunger to siphon water from a bucket or well, then pushed it forward while a third man aimed its stream at the fire.

The 1660s saw two major events that profoundly changed American history, and how fires were fought in colonial America.

The first occurred in New Amsterdam in 1664. For years, the Dutch colony's stern governor, Peter Stuyvesant, had imposed ever harsher regulations and ever higher taxes on the citizens, including a 1658 tax to pay for the addition of 250 new leather buckets for the city's fire protection. So when the English fleet, anchored off the southern tip of Manhattan (now the Battery), threatened to occupy New Amsterdam, the fed-up colonists put up no resistance and did not come to Stuyvesant's aid. The British took over without a fight, and renamed the colony New York. The new English governors continued Stuyvesant's strict fire regulations, and the city did not suffer a major fire until 1741, even though Boston had four Great Fires between 1654 and 1741.

The second event that shaped American firefighting history was the Great Fire of London on September 2, 1666, so it is perhaps appropriate that the New York Fire Department's library has an original newspaper account of that fire in its collection.

The Great London Fire of 1666 rampaged for five days. London was then the world's largest city, but by September 7, 13,000 homes, 87 churches, and countless other buildings lay in ruins. Destructive as it was, it brought with it two blessings. First, the fire wiped out London's rat-infested back alleys, which had spawned deadly plagues every 50 to 100 years since the Middle Ages; there would never be another great plague in London after the Great Fire. Second, in response to the Great Fire, English engineers developed hand-operated piston pumps that could be used to quell such fires in future.

Boston, the first American city to buy a fire engine, would also be first to buy one of the new English hand pumpers.

In 1672, another firefighting innovation was developed, this time in Holland. The van der Heijden brothers, Nicholas and Jan, developed a leather hose that could be placed between a fire engine and its nozzle, so the engine could be placed further from the fire and not be endangered should the fire spread. They also developed a suction pump that could draw water from a canal. It then used gravity to force the water through a canvas hose, elevated on wooden stands, down into the fire engine tub. This eliminated the need to hand-carry buckets of water from a water source to an engine.

In 1676, Boston suffered another Great Fire of its own. This one lasted but four hours before a heavy rain extinguished it. Fifty homes were destroyed, along with warehouse, stores, and Reverend Increase Mather's church.

But Boston citizens of 1676 had several advantages that their forerunners lacked in the 1653 Great Fire. Not only did they have the little Joseph Jynks fire engine imported from England in 1654, but they also had salvage bags. A salvage bag was a large sheet of canvas, usually made from an old ship's sail, folded double on itself, and a wooden rod joining the two ends. A crew of citizen volunteers could fill the salvage bag with personal belongings of someone whose home was in the path of a fire, and carry them to safety.

The only things that could not be moved

Canvas salvage bags let American colonists carry personal belongings out of homes that were in the path of a raging fire, and move these articles to safety. *Ed Hass collection*

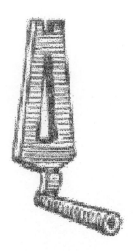

A bed key was a crude wooden wrench that colonial American fire wardens carried, to quickly disassemble beds for removal from houses threatened by fire. *Ed Hass collection*

in a salvage bag were the massive, heavy wooden beds common in colonial America. But American ingenuity had solved that problem, too: someone invented the bed key, a wooden wrench with a teardrop shaped hole in it. Some part of the hole was sure to fit over any size screw or dowel pin that joined the parts of the bed together. A bed key had a crank handle on one end to turn the screw or pin in or out of the bed, which made disassembling beds a matter of a very few minutes. In addition to their fire alarm rattles or hand bells, fire wardens, who patrolled cities like New York and Boston, now carried these bed keys as well.

The 1676 Boston fire was a first in American history: its cause was strongly believed to be arson. It also led to another American first: Boston city officials went to London to investigate the new hand pumpers, with one or two pump pistons worked by hand poles. These engines, carried on four men's shoulders or, in some cases, on crude wooden cartwheels, were just gaining popularity as fire protection in English towns. Back home in Boston, after much haggling, these officials finally placed an order for one of these engines, and it arrived in April, 1679. It was housed in a shed next to the prison on Court Street. Thomas Atkins, a carpenter, was appointed chief engineer in charge of this engine, and he had 12 assistants.

America's first true fire engine, manned by the first organized fire department, and quartered in the first firehouse, did not have to wait long to prove their effectiveness. In August, 1679, an arsonist lit up the Sign of the Three Mariners alehouse. Even with the new fire engine, however, the fire did not stop before it consumed seventy warehouses and eighty houses. Atkins and his crew of 12 were well paid for their backbreaking labor at this blaze. Another Boston firefighter, pinned under heavy beams at a fire on Fleet Street, was rewarded with £20, and a license to keep a public house and sell liquor.

In 1698, those inventive Dutch brothers, Nicholas and Jan van der Heijden, developed a way to attach a hose directly to the suction side of a fire engine pump. The pump could draw water directly from a canal into its own tub on each piston upstroke, forcing it out through the dis-

charge hose on the next down stroke. This eliminated the cumbersome canvas hoses, wooden elevating supports, and separate suction pumps of their 1672 method.

Following a fire of October 2, 1711, which destroyed over a hundred houses, and dozens of stores and churches, Boston ordered three more hand pumpers from England. By 1715, the city had six of these "English water engines," making it America's largest fire department at that time. Neither New York nor Philadelphia had yet purchased a single fire engine. Boston's six fire engines were housed in various sections of the city, and a militia officer was placed in charge of each engine. Each engine had a volunteer crew of 10, selected from among citizens who were considered "prudent persons of known fidelity."

The officers in charge of each Boston fire engine had authority to order all citizens to help out at fires, whether it was tearing down or blowing up houses, removing household items in salvage bags, or relieving exhausted firefighters on a hand engine's brakes. Those disobeying a direct order of one of these "Firewards" were fined 40 shillings, and the money was given to those who suffered the greatest loss in the fire. If a fire warden caught anyone looting during a fire, he had the authority to arrest the thief and inflict the harshest legal punishment.

To ensure that every citizen knew who the appointed Firewards were, each warden carried a staff signifying his high office. The wooden staff was 5ft long, painted red, and had a 6in brass spire at its head.

Although Boston had six fire engines, as many fire wardens, and 60 firefighters, the heavy damage done in so many fires over the previous century had Boston citizens concerned. So on September 30, 1718, a group of Boston home owners formed the Mutual Fire Society, a volunteer fire compa-

The fire warden's red staff, five feet long, was a badge of authority over all citizens in event of a fire. Boston used these "Fireward" staffs in the early 18th century. *Ed Hass collection*

ny of homeowners who mutually agreed to fight any fire that started in, or threatened to spread to, the house of any of the Society's members.

Boston's Mutual Fire Society was only the first in a long line of volunteer fire companies that would dominate not only American firefighting, but also politics and social life, for the next 150 years. Most prominent politicians of the late 18th and early 19th century were former volunteer firefighters, including George Washington and Benjamin Franklin. And in cities and towns all over Young America, the average citizen's social calendar was a whirlwind of firemen's picnics, parades, musters, and dances.

Colonial Boston's volunteer firefighters did not replace the existing engine companies, but rather supplemented them. When a church bell sounded a fire alarm, each volunteer grabbed his bucket, salvage bag, and bed key and ran to assist the regular firefighters. These few crude tools of the volunteers were each marked with the name of their fire company, and the individual name of the tool's owner. While the regular firemen were pumping the hand engine's brakes, these volunteers would carry valuables to safety in their salvage bags, dismantle and cart away beds, or refill the wooden tubs of the engines with buckets of water drawn from rivers, lakes, or wells. One of the volunteers would always stand guard outside of a burning house, to make sure that only town-appointed firefighters and members of the mutual fire societies, and not looters, were carrying away the homeowner's valuables.

One of colonial America's greatest innovators in firefighting came out of this fire-conscious environment: Benjamin Franklin, born in Boston in 1706. Just five years later, young Franklin witnessed what was, up to then, America's most spectacular conflagration: the Boston Towne House fire of 1711. He admired the work of both the town-appointed firemen, and the volunteers of the Mutual Fire Societies.

In 1723, Franklin moved from Boston to Philadelphia. Although founded in 1682, Philadelphia had yet to suffer a Great Fire, as Boston and New York had in their early history. In Philadelphia, as in other colonial cities, citizens were required to have brick, and not wood, chimneys, and to keep those chimneys clean when not in use. Any Philadelphia homeowner whose chimney was so dirty that it caught fire, was fined 40 shillings.

Philadelphians were also required to keep a bucket, and a long-handled swab, handy in case of fire, and were forbidden to store more than 6lb of gunpowder in any one building.

In 1718, Thomas Bickley, a resident of Philadelphia, somehow acquired a small hand pumper made in England. He sold it to the city for a mere 50 pounds, and it was the city's only fire engine when their first Great Fire struck homes and stores along Fishbourne's Wharf.

In the same year that Benjamin Franklin moved from Boston to Philadelphia, 1723, American inventors Selleck and Pennock improved on Dutch inventor Jan van der Heijden's 1672 leather hose with hand-stitched seams. Hand stitching was very time consuming, and tended to unravel, so instead, Selleck and Pennock used copper rivets to strengthen the seams of leather fire hoses. This innovation quickly proved popular not only in the colonies, but in Europe as well.

Leather hose still had one serious flaw, however. Like all leather, it dried out, becoming inflexible and even cracking, with age. To solve this problem, makers of leather hose recommended periodically treating the leather with grease. Firefighters disliked this, because the grease made the hose too slippery to maneuver at a fire, and the grease attracted soot from the fire. But it would be more than a century before a better alternative was available.

Two or more inventors often develop the same invention independently at the same time. And so it was with the first two fire engines built in America, both completed in January, 1733, one in Boston and the other in Philadelphia.

The Boston engine, designed and built by John and Thomas Hill of that city, was pulled by two horses, and demonstrated that it could throw a stream of water twelve feet into the air. The stream distance of the Philadelphia engine, built by Anthony Nicholls, was apparently not recorded, but it was noted that the stream shot higher than any of Philadelphia's London-built hand pumpers. These engines were too heavy, cumbersome, and expensive to be practical, and neither city purchased their hometown product, preferring to continue using the more advanced English hand pumpers.

Unlike Boston and Philadelphia, New York had largely escaped having its own Great Fire, despite having no fire engines. This was due largely to the strict building codes that Peter Stuyvesant had established, and which continued in force through the British taking over in 1664, the Dutch recapturing New York in 1673, and the British capturing it back again in 1674. But the city was growing, and in 1731, the city council ordered two hand pumpers from Richard Newsham of Cloth Fair, London, a manufacturer of pearl buttons.

Newsham's fire engine was so revolutionary that the British government awarded him a patent on it, on December 26, 1721. His first engine, built in 1720, consisted of two pump pistons, each 4.5in in diameter, with an 8.5in stroke (vertical motion). Iron rods tied these pistons to two smooth oak poles, called "brakes," extending 15ft along both sides of the engine. As one brake went up, the other went down, in see-saw fashion, moving one piston down and the other up. The up stroke of each piston drew water from the engine's wooden tub, and the down stroke forced it out through a "gooseneck" nozzle on top of the engine.

The Newsham engine discharged two quarts for each down stroke of the pump pistons, so with a crew of ten strong men pumping its brakes up and down at a steady 60 strokes per minute, it would deliver 30 gallons per minute. Compare that with today's fire engines, which discharge 2,000 gallons per minute. The engine's wooden tub could be refilled through a suction hose dropped into a river, lake, or cistern, with each upstroke of a piston drawing water into the tub. Or the tub could be refilled by pouring water directly into it from buckets.

The entire engine was mounted on four crude wooden wheels of the type used on ox-carts, not the fancy spoked wheels found on elegant carriages.

One Newsham engine, demonstrated in London, threw a stream of water over the top of a tower 165ft high, in view of several thousand spectators. These engines worked so well that at least one was still in active

firefighting service in England in 1865, 144 years after the original patent was granted.

The two Newsham engines for New York arrived on December 3, 1731, and were named Engine 1 and Engine 2. Both were kept in a shed behind City Hall, at Wall and Nassau streets. Peter Rutgers, a brewer and alderman, was overseer of both engines. With just these two engines, New York would keep from having its own Great Fire for another ten years.

In 1736, Benjamin Franklin established Philadelphia's first volunteer fire department, Union Fire Company. Its members

each had their own leather bucket and salvage bag, which they were required to bring to every fire. The members met once a month to discuss new ideas in preventing fires, and extinguishing those that did occur. The membership grew too big, and a second volunteer fire company was formed. Soon, Philadelphia had many volunteer fire companies, and nearly every property owner in that city was a member.

The two Newsham hand pumpers were New York's only fire protection on March 18, 1741, when a fire started on the roof of Lieutenant Governor George Clark's official residence, inside Fort George at the lower tip of Manhattan. Both engines were summoned, but the wind spread the fire until it not only burned down the governor's residence, but the chapel, secretary's office, and military barracks as well.

The lieutenant governor initially announced that a plumber had been repairing a gutter on the house, using a pan of coals to heat his soldering iron, when the strong wind blew a hot coal up onto the roof. But New Yorkers were not satisfied with that explanation, and began seeking another cause.

New York had been a center of the slave trade since the Dutch brought the first slaves over in 1628. A slave market had been established on Wall Street in 1709. By the time of the Great Fire of 1741, 2000 slaves lived in New York, one-fifth of the city's 10,000 population. The other 8,000 New Yorkers lived in constant fear of a slave rebellion, and some though the fire at the governor's mansion might have been part of a slave plot.

The city had suffered many small chimney fires over the years, and all were quickly extinguished. Such a fire occurred in the home of Captain Warren just one week after the Fort George fire. Over the next few weeks, every fire was seen as part of this slave conspiracy: a cow stable burned, a kitchen caught fire, and a careless smoker ignited a warehouse.

Mary Burton, a 16-year-old indentured servant to the John Hughson, owner of a working-class rum joint, had accused two of Hughson's slaves, Caesar and Prince, of stealing a few dollars worth of linen, and they were in jail for that unproven crime. When Lieutenant Governor Clark offered a reward for information about the recent fire in his residence, young Burton spun fanciful tales of how Hughson and his daughter Sarah, along with the two imprisoned slaves and a local prostitute named Peggy Kerry, had plotted to burn down the entire city. A slave who had been overheard singing about fire was beaten into confessing to a role in that plot, and Sarah Hughson's fortune-teller friend testified to a supernatural knowledge of the plot. Within a month, New York's small prison was overcrowded with accused conspirators in the plot to burn the city down.

So when a real arsonist ignited seven barns across the river in Hackensack, NJ, on May 1, 1741, it was decided to end the slave-revolt conspiracy, and all accused were summarily hanged. John Hughson denied any plot right to the end, and prostitute Peggy Kerry admitted on the gallows that her tales of a slave revolt were her own invention.

Small fires in New York did not stop with these executions, of course, and other conspirators were sought. When John Ury was arrested on charges of being a Roman Catholic priest, Mary Burton accused him of being a conspirator in the plot to burn down New York. She was never asked why she had not mentioned this in the original trial, and with no evidence against him, Ury was found guilty and hanged.

After the next few small fires, Mary Burton upped the ante, accusing some prominent New Yorkers, including a magistrate, of being involved in the slave plot. At last, New Yorkers figured out that Burton's testimony had been entirely a fabrication of her evil mind. But it was two late to save the 154 jailed slaves, the 13 more burned at the stake, 18 hanged, and 70 more transported to prisons. The realization that there was no plot to burn down New York failed to save 20 free men who were jailed, four more who were jailed, and eight more transported to prisons. The bodies of John Hughson and the slave Caesar were left outside on exhibit all summer, even after it was obvious they had done nothing wrong. And the following March, Mary Burton, whose testimony was now known to be lies, nevertheless received her reward of £170 for finding those who set fire to Lieutenant Governor Clark's house!

In 1743, Thomas Lote of New York completed the first successful fire engine designed and built in America. It became Engine 3 in the fledgling New York fire department. While the two English-built Newsham hand pumpers had their brakes along both sides of the engine, the American-made Lote engine's brakes were at front and rear in a design that would later be known as the "Philadelphia style." Besides the brass nozzle on top, this all-wood engine had all edges trimmed in brass, earning it the nickname "Old Brass Backs."

Shortly after purchasing the Lote engine, New York received another hand pumper, previously ordered from England. It became Engine 4. Lote is not known to have ever made any more fire engines after the one for New York in 1743, and fire engine manufacturing remained an English, not an American, industry.

About 1750, another firefighting innovation swept colonial American cities: the idea of motivating volunteer firefighters to run to a fire as quickly as possible. A cash reward from the town treasury, usually about £5, was offered to the first fire company to arrive on the scene. Insurance companies sweetened the pot, especially if the first fire com-

New York City Fire Museum (278 Spring Street) owns this mid-18th century hand pumper.

took in from subscribers. What little fire insurance there was in colonial America before Franklin's 1752 innovation, was underwritten by companies in the mother country, England.

In 1760, Boston, devastated by Great Fires in 1653, 1673, 1679, and 1711, suffered its fifth Great Fire. This one destroyed 174 houses and 175 stores and warehouses. Loss was estimated at £100,000 sterling, or about $2million in 1998 dollars.

In 1768, Richard Mason of Philadelphia established the first American fire engine manufacturing company. His first fire engine, with brakes on the front and rear instead of the sides, went in service at Northern Liberty Fire Company of Philadelphia. Mason manufactured dozens more of these fire engines over the next 40 years. While Newsham engines had the tower for pistons and brake levers at the rear, the Mason engines placed them squarely in the center, distributing the engine's weight more evenly on all four wheels. This balance was crucial, because the wheels of a Mason engine were smaller than those on a Newsham. Mason engines had separate refill areas for pouring water from buckets, one for the piston at the front of the tub, and one for the rear piston. One Richard Mason hand pumper was credited with throwing a stream of water to a height of 75ft.

The Hibernia Fire Company of Philadelphia had purchased a Newsham hand pumper in 1752, and by 1790, it was worn out after 38 years of continuous service. So they replaced it with a new Richard Mason hand pumper in 1790, for £160. The low price and local manufacture made the Mason engine more attractive than English fire engines in many communities of the late colonial era.

Several Richard Mason hand pumpers survive today, including one that, incredibly, had been stored in a Philadelphia attic for longer than anyone could remember, and was rescued by a Philadelphia firefighter when that attic caught fire in the late 1960s. This firefighter was friends with a local doc-

pany on the scene answered a call to a fire at a home displaying that company's fire insurance mark.

This financial incentive worked well for a few years, but by the mid-19th century, the competition to be first on the scene spawned fierce rivalries between a city's various fire companies, often deteriorating into bloody brawls at a fire scene. As we already saw in the introduction of this book, one such riot, in Cincinnati in 1851, paved the way for the introduction of paid fire departments and steam fire engines.

In 1752, Benjamin Franklin, who had established Philadelphia's first volunteer fire company 16 years earlier, founded America's first successful fire insurance company. The Philadelphia Contributorship for the Assurance of Houses from Loss by Fire pooled the subscription fees paid by its customers, and paid out to those who suffered financial loss from a fire. Subscribers were given a "Fire Mark" plaque with the company's "Hand-in-Hand" logo, to nail to the outside of their homes, not only as proof of insurance, but also as good publicity for the insurance company.

As early as 1735, a fire insurance company called the Friendly Society had started in Charleston, SC, but that company quickly failed from paying out more in losses than it tor who collects fire engines, and the Mason hand pumper was still in that doctor's barn in 1990. Several museums in the eastern U.S. also have Richard Mason hand pumpers.

In 1784, inspired by Benjamin Franklin's success in the fire insurance business, another fire insurance company, Mutual Assurance Company, sprang up in Philadelphia. While Franklin had used the Hand-in-Hand logo on a wooden fire mark, this new company outdid him, using a green tree cast in lead, mounted on a wooden shield. The stage was set for rivalries between fire insurance companies, and between the volunteer fire companies that each insurer sponsored.

Philadelphia was fast replacing London as the center where improvements in fire suppression and loss reduction developed. In 1794, another Philadelphian, Pat Lyon, developed a much larger version of the end-stroke hand pumpers that Richard Mason had built. In 1835, another Philadelphia hand engine builder, John Agnew, improved on Pat Lyon's design, becoming the premier builder of "Philadelphia style" fire engines until 1860.

While all of these improvements were beginning to come out of Philadelphia, New York, and Boston, another change was in the wind in those cities: rebellion against the repressive tactics and stiff taxes that the English crown imposed on the colonists. Further fueling their discontent, these same overburdened colonists were not permitted representatives to speak for their interests in the English parliament.

The unrest culminated in the signing, at Philadelphia on July 4, 1776, of a Declaration of Independence from England. This declaration triggered a decade-long war of the colonial rebels against the British army. When the colonists finally obtained their victory, they still had to form their own government, and draw up a Constitution by which the new government would operate. And so it is not until 1790, 14 years after the Declaration was signed, that the colonial era, and this chapter of American firefighting history, drew to a close.

The American colonial era would suffer two more Great Fires before drawing to a close, however. This time, the fires were in neither an English nor a Dutch colony, but a French one: New Orleans, founded in 1718. That city's first Great Fire, on March 21, 1788, began when an overturned candle ignited some drapes. By the time it ended five hours later, 816 buildings lay in ruins, in a city with a population of 5,000.

Following the 1788 fire, citizens of New Orleans rebuilt their city, and established four volunteer fire companies to prevent a repeat of that disaster. Each company, in the four districts of the city, was equipped with buckets and a hand pumper. But even these precautions did not prevent a second Great Fire from destroying 16 city blocks of New Orleans in 1794.

This well-preserved Richard Mason hand pumper, built about 1785, is one of several in museums in the eastern U.S. Mason was the first American to establish a factory for the manufacture of fire engines. *Ed Hass collection*

CHAPTER TWO

Fire Equipment in Young America 1791-1852

Although America declared its independence from England in 1776, it would take a long and bloody Revolutionary War to make the desire for independence a reality. When that war ended, America had to form a federal government and draft a Constitution containing the basic governing principles of the new nation. So the Colonial era did not really end on July 4, 1776, but in 1790. The new nation was brimming with inventive geniuses, and some of them applied their ideas to improving the fire service.

One idea was to protect the firefighter's skin and clothing from flames, smoke, and water. In 1794, firefighters of Assistance Fire Company in Philadelphia began wearing protective capes treated with oil, clay, and pigments.

These "oil cloth" capes repelled water nicely, and kept sooty smoke from staining the clothes underneath, but offered no protection from heat and flames. In fact, since the capes were often lettered with the fire company's name and founding date, using highly-flammable paints, they were more likely to subject their wearers to heat blisters and burns.

A few of these firemen's capes survive in museums, and fewer still in the hands of collectors. If you can even find an original early-American fireman's cape for sale in good condition, acquiring it is likely to set you back anywhere from $3,000 to $10,000.

America took five years from the end of its revolution, to inaugurate General George Washington as its first president. These were years of defining a distinctly American character and culture, leaving behind its English, German, Dutch, and French roots. The world's newest nation promised its citizens an unprecedented level of freedom, and required of them an unprecedented level of participation in the new nation's government.

Diversity was already becoming an American trait. Early sessions of the new congress could not decide on an official language for the new nation. In the cities, with its brick buildings and large factories, English was the primary language, but sometimes spoken with a Scottish or Irish accent. In the villages, where farms and mills worked cooperatively, and church spires towered over the landscape, you were as likely to hear German or Dutch spoken as you were to hear English. And in the cabins at the edges of the unexplored wilderness, you would find French-speaking fur trappers, English and German farmers, Scottish and Irish adventurers, and Indian tribes speaking a bewildering variety of Native American languages.

And so the first Congress took what would become a distinctively American action: they decided not to decide. Over 200 years later, America still has no official language, and you are likely to hear Spanish, Italian, Chinese, Japanese, or Vietnamese in the nation's cities, suburbs,

Fire insurance companies often organized and operated early American volunteer fire companies. If you subscribed to that company's insurance, they placed a fire insurance mark on your house, and the volunteers would extinguish any fire in your house. But if you lacked an insurance mark, or had a mark from a rival insurance company, that company's firefighters often let your house burn. The New York City Fire Museum displays this collection of late 18th century Fire Insurance marks. Two or four clasping hands, trees, and eagles were common themes on these marks. Notice that most marks included the customer number of the insurance subscriber, but one bears the year for which the insurance was paid: 1794.

and rural areas, as well as the many languages of America's founding citizens.

Young America had many tasks on its hands. Not only did it need to rebuild a war-damaged nation, and establish new laws and new governing principles, but it also had to retrain its many soldiers for productive civilian life. Many found a home in the fire service, which was run much like the type of military organization to which they had belonged in the Revolutionary War. As firefighters, these former soldiers could still feel that they were putting their lives on the line to make a vital contribution to the nation. In combating flames and smoke, they could feel hostility toward an enemy, and the excitement of victory over that enemy. And the comradeship among firefighters was much like that among soldiers.

The volunteer fire companies of the post-Revolution era had much the same function as today's veteran organizations. The engine house was their clubroom, and the weekly meetings were a chance for the soldiers-turned-firefighters to sit down together to a meal of fish chowder or mutton pie, to drink together, to swap tall tales and catch up on the latest developments in each other's personal lives. They would lavish as much care on their engine house and their fire engine, as they had on their military barracks and their muskets in the Revolutionary War. At least once a week, and always after a fire, they would wash and polish their engine, oil the wheels and the moving parts of the pump, and thoroughly dry the owing rope, leather hose, and nozzles. They would even spend their free Saturday afternoons experi-

With the heady feel of victory over the British army, and the establishment of the first modern nation based on democratic principles, Americans of the late 18th century began doing everything bigger and better, even fighting man's oldest enemy, fire. Bigger than the English engines of Richard Newsham or the Philadelphia engines of Richard Mason, post-Revolution hand pumpers even featured elegant spoked carriage wheels. These two well-preserved hand pumpers at Millville, NJ, are typical of the era. The engine at left dates from 1795, and at right from 1798. *Ed Hass*

menting with the best pace at the brakes, to get the highest water volume or pressure out of the pump. This experimentation was much like the way military strategists experimented with he best angles for the most effective trajectories from different types of guns and cannons.

One example of the transition from being an English colony to a separate country, was the reorganization of the New York fire department in 1786. The city's 15 engine companies and two hook-and-ladder companies were placed under the command of five city-appointed engineers. Volunteers were sought from among "strong, able, discreet, honest, and sober me" until the city's volunteer firefighting force increased to 300 members.

As American cities (such as New York, Boston, and Philadelphia) grew larger, so did the fire engines that protected them. From the 4.5 x 8 pump of the English Newsham and American Mason engines, the typical American-made hand pumper of the post-Revolution era had 6.5 x 9 cylinders. The number of men required to operate these pumpers grew from 10 to 15 or even 20. With the increase in the size of fire engines, came new manufacturers of these large engines: William C. Hunneman of Boston (1792), James Smith of New York (1813), and Lysander Button of Seneca Falls, NY (1835).

Fire engines were becoming so big that they could not easily maneuver, once placed at a fire. This led to another American innovation: threading the pump inlets and outlets, and the hose couplings, so that hose could be screwed onto the pump. Rather than moving the engine, firefighters cold quickly and easily couple extra lengths of hose to an existing hose line, if the wind carried the fire away from the engine, or if the firefighters need to advance further inside a warehouse or factory.

As cities spread over larger areas, extra engine houses were built, and more engines purchased. But the pace of adding to the fire department usually lagged behind the growth in population and area. This meant that firefighters often had to cover more territory than in previous generations. And the territory they covered was more densely populated, and therefore noisier. The traditional watchman's rattle could not carry a noise over the increasing volume of din, or even the increasing distances between a fire and the firehouse. Church bells were briefly tried and abandoned, because citizens were confused whether the bells meant to run to a fire, or to head to the church for a religious service, especially if the fire alarm sounded on the church bells on a Sunday morning.

A new type of brass bell, with a much deeper, richer, and more urgent tone, was developed for the fire service. To all but the tone deaf, there was no longer any confusion between the desperate jangling of the fire bells, and the more soothing melody of a church bell. American author Edgar Allen

Early American fire-fighters took great pride in their hand-operated fire engines, and would decorate them elaborately. They often hired the most renowned artists of the day to paint a landscape, or a portrait of a famous person, on the sides of their engines. The quality of decoration was almost more important to early American firefighters than the effectiveness of their engines and hand tools in extinguishing fires. Bates and Jeffers built this fancy engine, the "Pacific," for Pawtucket, RI, in 1844. It is now at the Hall of Flame, America's largest fire museum, in Phoenix, AZ.

FIRE EQUIPMENT

Poe summed up this difference in his circa-1845 poem, The Bells. First, church bells announce the joy of a church wedding:

Hear the mellow wedding bells
Golden bells
What a world of happiness
Their harmony foretells
Through the balmy air of night
How they ring out their delight
Through the dances and the yells
And the rapture that impels
How it swells
How it dwells
On the future
How it tells
From the swinging and the ringing of the molten golden bells
Of the bells, bells, bells, bells, bells, bells, bells
Of the rhyming and the chiming of the bells.

But here is how Poe describes the American fire alarm bell in the next verse:

Hear the loud alarum bells
Brazen bells
What a tale of terror now
Their turbulency tells
Much too horrified to speak

Oh, they can only shriek
For all the ears to know
How the danger ebbs and flows
Leaping higher, higher, higher
With a desperate desire
In a clamorous appealing to the mercy of the fire
With the bells, bells, bells, bells, bells, bells, bells
With the clamor and the clanging of the bells.

Standards for the manufacture of leather fire hose were developed in 1817. Each piece of leather weighed 20 to 22lb. Their seams were double riveted, using rivets made from Number 8 copper wire, spaced 22 rivets to the linear foot. Splices between sections of leather in a single leather hose consisted of 13 rivets, all made of No. 7 copper wire. A 50ft length of leather fire hose weighed 64lb, without the hose couplings. Strict testing was required: the hose had to withstand a pressure of 200lb per square inch.

Rubber fire hose was developed in 1827. But, as with many innovations, this one was adopted slowly, not becoming common in American fire stations until the 1870s. The last copper-riveted fire hose in regular fire-fighting use in America was finally retired about 1900.

A single 50ft length of rubber fire hose, wrapped in two woven cotton jackets, weighed 50lb, including the heavy brass couplings at both ends. They could withstand a pressure of 400lb per square inch. Not only were these hoses lighter and stronger than leather, but they did not dry out and crack, or require oil to keep them flexible. They did, however, mildew if not dried immediately after each use. The severe abrasion of dragging the cotton against brick and stone pavements sometimes caused these hoses to burst, even under moderate pressures.

Yankee ingenuity solved at least one problem with this new hose: the tendency to burst when abraded. Various inventors developed "hose sleeves" or "hose jackets," clamps of iron, steel, or other sturdy metal, that could be placed over the rupture. Water flow continued uninterrupted through the hose, until the fire was extinguished.

Another improvement in early American firefighter's tools was in the area of fire helmets. By the early 1800s, the top-hat style helmet had given way to one with a curved top, more nearly matching the shape of a human head, and therefore offering better skull protection from falling ceilings and other hazards of fighting fires inside buildings. But as with many other Yankee innovations, this one was not quickly nor universally accepted. The pre-Revolutionary stovepipe felt helmet did not fall entirely out of favor with American firefighters until the Civil War in the 1860s.

In 1824, Matthew DuBois added a metal wire framework to the helmet, preventing the leather from warping in the heat of a fire.

The fire helmet underwent one more improvement in 1836, when New Yorker Henry Gretacap noticed that water from fire hoses ran off the flat brim of a firefighter's helmet, down his neck, and inside his coat or shirt. He decided to bend the back of the brim downward, to let water flow safely away from a firefighter's neck and clothing. By almost pure serendipity, this bent brim also protected a firefighter's neck from the scorching heat of flames at a fire. Gretacap also noticed that the top of the leather frontpiece, identifying the wearer's fire company, would warp in the heat of a fire. So he attached a brass eagle to the front of the helmet, with its beak holding the top edge of the frontpiece. In 1868, Gretacap sold his thriving fire helmet business to Jasper and Henry Cairns of New York. Today, Cairns & Brother, now in Clifton, NJ, still makes the same basic "New Yorker" leather fire helmet.

In the Colonial era, Hook and Ladder companies mostly consisted of a fire house equipped with ladders, hooks for pulling down chimney-tops and burning roofs, and axes for chopping holes to ventilate smoke

Left: As America's cities grew in population and area, volunteer firefighters found clever ways to create a loud enough noise to summon them to fires. Many small towns kept a huge steel locomotive tire, and a mallet heavy enough to strike it, in the town square. Cincinnati kept a giant drum in a tall tower, where its thunder could be heard at great distance. McNeely & Company of West Troy, NY, cast this huge brass fire bell for the same purpose, about 1850.

from buildings. Usually, the members of the Hook & Ladder Company carried these tools from the firehouse to a fire. If they were lucky, they had a cart on which to pile the ladders for transit to and from fires. In Colonial New York, the hook and ladder company did not even have a fire house: ladder were hung along a fence in City Hall Park and hand carried from there to a fire.

After the Revolution, special carts, or trucks were designed specifically for carrying heavy wood ladders and other 'truckee' tools to fires. Many of these ladder trucks steered at both the front and the rear, for easier turning around tight corners.

The first vehicle designed specifically as a fire department ladder truck was built in Philadelphia in 1799. It had separate racks for each ladder, and rollers for easier loading and unloading of the ladders. Its longest ladder measured 40ft long. It also carried

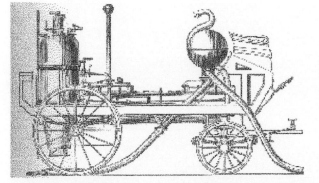

long wooden poles with hooks on the end, so was truly the first Hook and Ladder Truck.

In 1847, Dudley Farnum and Franklin Ransom of New York patented a novel type of hand pumper, called a"rowing engine." Its pump was in the center of the engine, with two rows of bench seats at the front, and two more at the back. All four seats faced inward, toward the pump. Instead of standing along both sides of the engine and pumping the brake levers up and down, firefighters sat atop the engine's deck and rowed as if moving a boat across a lake. The inventors claimed that this method exerted more force, with less strain on the operators. This engine never proved popular, and it would probably have been forgotten had not Abel Shawk of Cincinnati chosen a massive Farnum hand-engine pump for his prototype steam fire engine in 1851.

Which brings me to the invention (steam fire engines) and the event (introduction of paid, professional fire departments) that closes the era of the hand pumper operated by volunteer firemen.

Hydraulic Engineer George Braithwaite and 26-year-old, Swedish-born inventor Captain John Ericsson, both of London, England, built the world's first recorded steam fire engine in 1829. This was just 26 years after another English engineer, Richard Trevithick, developed the first steam railroad locomotive, for carrying coal out of mines. And it was only four years since Stephenson's "Rocket" engine adapted the

railroad locomotive for hauling passengers.

Braithwaite and Ericsson's first steam fire engine, the Novelty, weighed 4,500lb. Its pump pistons were 6.5in in diameter, and its steam cylinders 7.5in in diameter. It could pump 250 gallons of water per minute, and throw a stream to a height of 90ft. The inventors used it at several fires in London, and repeatedly proved its effectiveness, but tradition-bound London firefighters and insurance companies did not want this smoke-belching engine. Braithwaite and Ericsson did sell this engine, and two more like it, to remote English villages desperate for any form of fire protection.

Finding no interest in his home country, Braithwaite in 1832 built his fourth steam fire engine, for the King of Prussia, to protect Berlin's public buildings. Named Comet, it was larger than Novelty, with 12 x 14 steam cylinders powering 10.5 x 14 pump pistons. Starting from a cold boiler, it could raise 70lb of steam in 20 minutes (a fire could destroy a lot while the engineer waited 20 minutes). The pump pistons moved up and down at 18-25 strokes per minute, and it handled up to four hose lines at once. With one line, and a 1.25in nozzle, it threw water 115ft in the air.

In 1833, Braithwaite and Ericsson built an experimental steam fire engine, fitted with the boiler and propulsion of a railroad locomotive. This, too, found no reception in London, and was apparently dismantled and never sold.

In 1839, Captain John Ericsson, a decade older and considerably wiser, gave up trying to sell his innovations to tradition-bound Englishmen, and moved to America, where his inventions might be better received. Within a year after his arrival, the stage was set for the development of steam fire engines on the western side of the Atlantic.

The winter of 1839-1840 saw a rash of fires in New York. Alarmed citizens and fire insurance underwriters were ready for a more efficient way to extinguish fires. Ericsson gleefully dusted off the plans for the steam fire engine he and Braithwaite had developed 11 years earlier. Perhaps not trusting a foreign engineer who had an odd accent, the insurance underwriters asked New York's mechanical engineering genius, Paul Rapsey Hodge, to build them such an engine as Ericsson proposed.

Hodge began work on his steam fire engine at his Laight Street shop on December 12, 1840. At 4:00pm on Saturday, March 27, 1841, the first steam fire engine ever built in America had its first public demonstration, in front of New York's city hall. At 120 revolutions per minute, it could by far out-pump the hand engines, and it never fatigued as firefighters did. Its stream reached a height of 166ft, which was cause for much booing and hissing from the volunteer firemen, who feared that such an engine would dramatically change their rowdy and politically powerful lifestyles.

The Hodge steam fire engine, America's first, was 13ft 6in long, and weighed close to 8 tons. Much of this weight was due to the self-propulsion mechanism, as on locomotives. Its steam cylinders were 9.5 x 14in, and its pump pistons were 8.25 x 14in.

Hodge somehow persuaded George W. Lane, foreman of New York's Pearl Hose Company Number 28, to put this engine in service. But the firefighters were so determined not to like this engine that they found fault with every feature, even those that could be improved with minor adjustments. The insurance companies realized that the open hostility of the firefighters was liable to cost more damage than the steamer prevented, and after less than six months, they quietly sold Hodge's engine to one Mr. Bloomer, who used it as a stationary engine to power his packing-box factory.

The insurance companies had not been the only ones concerned about New York's high fire loss in the winter of 1839-1840. While the insurance underwriters were busy having Hodge build them a steam fire engine, the Mechanics Institute of New York City in 1840 offered a gold medal for the best steam fire engine design.

In 1841, John Ericsson drew up plans for a steam fire engine that was a vast improvement on the five that he and Braithwaite had built in London between 1829 and 1833. Only two other designs were submitted, and as Ericsson was the only one of the three designers who had actual experience building steamers, it is not surprising that his design won the gold medal.

The engine that Ericsson proposed would throw 3,000lb (about 360 gallons) of water per minute, through a 1.5in nozzle, to a height of 15ft. Its boiler would contain 27 tubes, each 1.5in in diameter, reaching a working head of steam from cold water in just 10 minutes (double the speed of his 1829 engine).

Apparently satisfied to win the gold medal, Ericsson never actually built or sold a steam fire engine in America. His most

famous application of steam power was not until more than 20 years in the future. Ericsson designed and built the ironclad steamship Monitor for the United States Navy during the Civil War, and it defeated the Confederate Navy's previously invincible ironclad, Merrimac.

Unlike John Ericsson, one of the other two contestants at New York in 1841 did not give up on introducing steam power to America's fire service. Abel Shawk of Cincinnati was a locksmith by vocation, but a steam enthusiast, tinkerer, and dreamer by avocation. Returning to Cincinnati from New York after losing the 1841 competition, he acquired a 20-year-old Buchanan steam boiler, dismantled it, and began experimenting with ways to make boilers smaller, lighter, and faster, so they would be practical for use in a steam fire engine. After 10 years of spare-time tinkering, he finally had a practical boiler in 1851.

Success is part being at the right place at the right time, and part being shrewd enough to grasp opportunities that come your way. Soon after Shawk perfected his improved Buchanan boiler, several hundred Cincinnati firefighters started a bloody and destructive riot in September, 1851. Citizens and councilmen began to push for ways to reform the fire department. One such councilman, Miles Greenwood, was a former volunteer firefighter, and owned the foundry where Shawk had just completed his prototype steam fire engine boiler.

Another fortunate circumstance was that another Cincinnati inventor, Alexander B. "Moses" Latta, was not having a great success in his steam railroad locomotive business. Latta was thus receptive to the idea of making a steam engine to convert power from Shawk's boiler into the drive force for

the pump from a Farnum "rowing" style hand pumper. He even built a locomotive-style propulsion mechanism to move the fire engine down the street under its own power.

One final fortuitous circumstance was that, unlike his counterparts in other cities, the Chief Engineer of Cincinnati's volunteer fire companies was receptive to change. In fact, Chief R.G. Bray so liked the steam fire engine, that he eventually contributed half the cost of its development from his own pocket.

Latta & Shawk's 1852 prototype steam fire engine was so impressive, that the city asked them to build one to their specifications. The prototype was dismantled and never sold, and the Cincinnati engine was completed in December, 1852.

Councilmen Miles Greenwood, Joseph Ross, Jacob Piatt, and others did not want to entrust the steamer to the volunteers. After all, volunteers had defeated the steamer's introduction in London and New York. And so it happened that on January 1, 1853, Cincinnati established the first fully-paid, professionally-trained fire department in America, operating the first successful steam fire engine in America. This first Latta & Shawk engine was named Uncle Joe Ross.

Before the Latta & Shawk prototype engine was completed, one more inventor would try, and fail, to introduce steam fire engines to America's fire service. In 1851, William Lay of Philadelphia drew plans for a steam fire engine, but finding Philadelphia's firefighters as opposed to stem engines as those in London and New York were, Lay never built his engine.

What made Cincinnati's story different from those of London, New York, and Philadelphia, was that the steam fire engine had popular support there. In fact, the 1853 steamer was so popular, that in 1854, the citizens donated a second Latta & Shawk steam fire engine (actually their third engine, counting the dismantled prototype), and a fire station to house it. Assigned to Fire Company Number 3, Cincinnati's second steamer was appropriately named Citizen's Gift, a name that Cincinnati Engine 3 proudly continued for nearly a century.

In 1855, Latta and Shawk dissolved their partnership, and each established their own firm for manufacturing steam fire engines. Shawk would only build five steamers before giving up. But Latta would build about 30 steam fire engines.

On March 1, 1857, Louisville, KY, became the second American city to organize a fully-paid, professionally-trained fire department. Like Cincinnati, Louisville built its new fire department around the use of Latta steam fire engines. St. Louis, MO, quickly followed the examples of Cincinnati and Louisville, establishing its paid fire department on September 14, 1857. They, too, used Latta steam fire engines as the catalyst for this change.

In the older cities on the east coast, such as New York, Philadelphia, and Boston, there was more of a desire to stick to tradition than in the newer cities of the midwest. And in those cities, the firefighters had amassed tremendous political power, so nobody wanted to cross them by introducing steamers. Philadelphia briefly tried, then abandoned, a Shawk steam fire engine in 1855. New York briefly tried one or two locally-built steam fire engines in the late 1850s.

But as America headed into the Civil War in 1861, the firefighters in eastern cities were still mostly volunteers, and their equipment was still mostly old-fashioned and hand-operated. Only after four years of bloody Civil War (1861-1865) would the fire service revolution, begun in Cincinnati in 1852, sweep the rest of the nation as it struggled to rebuild itself and grow into a world power.

Left: Close-up detail of the side panel of the *Phoenix*. *Ed Hass*

Below: John Agnew of Philadelphia, PA, built this double-deck, end-stroke hand pumper, the *Phoenix*, in 1825. This "Philadelphia-style" engine allowed firefighters on the ground and others on the engine's deck to operate the pump's front and rear "brake" levers. It now belongs to the fire museum at Benicia, CA. *Ed Hass*

CHAPTER THREE

Fire Equipment becomes Mechanized 1853-1880

With the success of the first Latta & Shawk steam fire engines in Cincinnati in 1853, other cities gradually followed suit. Philadelphia bought a steamer from Abel Shawk of Cincinnati in 1855, placed it in service in January, 1856, and facing bitter opposition from the volunteer firemen, sold it to a junkyard the following year.

Opposition to the steamer was even worse in Boston. Their Latta steamer, purchased in March, 1855, sat idle at a $1 million fire that swept large sections of Boston on April 29, 1855. Although the steamer had proven it could do the work of ten hand pumpers, the stubborn volunteers refused to use it. In November, 1856, on a routine practice drill, the top of the pump blew off violently, and it was later found that saboteurs had loosened all the bolts holding the pump together, so that the tremendous water pressure blew the pump apart. The $12,000 steamer was junked without ever having fought a fire.

Meanwhile, Cincinnati was using its original steamers of 1853 and 1854 so much, that they bought five more in 1855, another in 1858, and four more in 1861, just before the outbreak of Civil War.

St. Louis, MO, bought its first steamer, from Abel Shawk of Cincinnati, in October, 1855, and a second Shawk just four months later. A Latta steamer in 1856, three more Lattas in 1857, and one in 1858 followed this. St. Louis organized its fully-paid fire department on Sept. 4, 1857.

Louisville, KY, started its paid fire department with one Latta steamer on March 1, 1857, and followed up with four more Latta engines in 1858. Toledo, OH, and Indianapolis, IN, both bought Latta steamers in 1860. Meanwhile, in Seneca Falls, NY, Horace Silsby began building steamers for east coast fire departments in 1856. By the time his company merged into American Fire Engine Company in 1892, he had built over 1,000 steam fire engines.

Lee & Larned of New York briefly built steam fire engines, from 1856 to 1863. Likewise, Reanie & Neafie of Philadelphia entered the steamer business in 1857, and gave it up in 1870 after building about 25 engines, mostly for Philadelphia, Baltimore, and Washington. They also sold steamers to smaller cities, such as Providence, RI; Augusta, GA; Madison, IN; Nashville, TN; New Orleans, LA; San Francisco, CA; and Mobile, AL. They even built one for Cuba. Incredibly, the first two Reanie & Neafie engines built both survive today, one at the Smithsonian in Washington, DC, and the other at CIGNA Insurance Company's fire museum in Philadelphia, PA.

In 1859, Amoskeag of Manchester, NH, a highly successful builder of steam railroad locomotives, began making steam fire engines. By the time Amoskeag ceased steamer production in 1913, they had built over 800 of these engines.

In August, 1854, Latta hired a young apprentice, Chris Ahrens, who had settled in

Cincinnati after arriving from Germany less than a year earlier. Latta retired from the fire engine business in 1863, after building about 30 steamers. He sold his company to Lane & Bodley, a local machine shop, and Chris Ahrens became L&B's superintendent of fire engine construction. In 1868, Ahrens bought out the Latta steam fire engine business, renaming it C. Ahrens & Co., and later Ahrens Manufacturing Co. He set about redesigning the pump, boiler, and chassis, and by the 1880s, his engines held every world's record for steam fire engines: fastest to generate steam and throw water, greatest volume and pressure of water, and farthest and highest fire streams. By the time his company merged into American Fire Engine Co. in 1892, he had built about 400 steamers.

In 1862, Clapp & Jones of Hudson, NY, and L. Button & Sons of Waterford, NY, entered the steam fire engine business as well. Button had been a successful builder of hand pumpers. Both firms did well, building two or three hundred steamers each, before both merged into American Fire Engine Company in 1892.

With the growing popularity of the steamer, came a host of specialized tools to operate and service them, and the hose wagons and hose reels that accompanied them to fires. This chapter will examine these tools, the types of apparatus that carried them, and some of the more spectacular fires where these "steamer era" tools were used.

Since America won its independence from England, it had been sharply divided on the question of just who was to have freedom in this new land of the free. As defined by the nation's founding fathers, only males of northern European ancestry, above the age of 21, and owning their own land, were eligible to participate in this freedom. All others, whether they were female, black, native Americans, or Asian, did not fully enjoy the rights and responsibilities of citizenship.

In the new Kansas territory in the 1840s and 1850s, many a bloody battle was fought over whether slavery would be legal there. Abolitionists in Boston published newspapers decrying the African slave trade introduced to America by the Dutch, carried on by the English, and still permitted in the

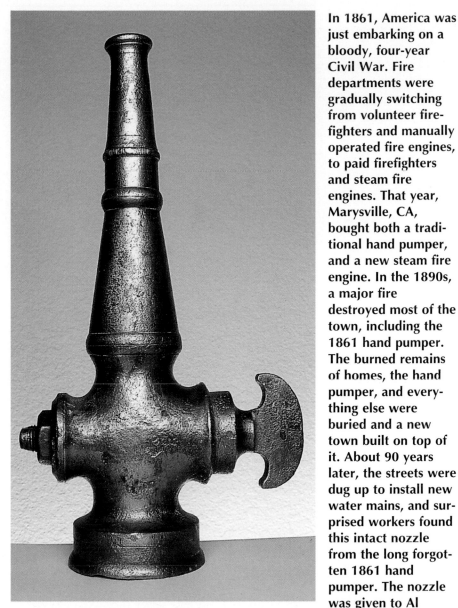

allegedly free southern states. In 1859, at Harper's Ferry, VA, fiery preacher John Brown led an unsuccessful slave revolt that found him and his followers holed up in that center of 19th century social activity, a volunteer fire station, where all perished at the hands of Federal troops.

Abraham Lincoln was not a rabid abolitionist, but he had publicly stated his dislike for the institution of slavery. As a result, southern politicians, whose fortunes had been made from the crops hand picked by such slaves, did not much care for Honest Abe. Lincoln's election as President of the United States in 1860, and inauguration in early 1861, stirred the anger of southern leaders. They declared their intention to form a separate nation, the Confederate States of America (CSA). It all came to a head when CSA soldiers fired upon, and captured, Fort Sumter in Chaleston, SC, lowering the U.S. flag and raising the CSA's

In 1861, America was just embarking on a bloody, four-year Civil War. Fire departments were gradually switching from volunteer fire-fighters and manually operated fire engines, to paid firefighters and steam fire engines. That year, Marysville, CA, bought both a traditional hand pumper, and a new steam fire engine. In the 1890s, a major fire destroyed most of the town, including the 1861 hand pumper. The burned remains of homes, the hand pumper, and everything else were buried and a new town built on top of it. About 90 years later, the streets were dug up to install new water mains, and surprised workers found this intact nozzle from the long forgotten 1861 hand pumper. The nozzle was given to Al Mazzerolle of Marysville Fire Dept., who sold his collection in 1992 to the author.

"Stars and Bars" flag. The American civil war was on. It would drag on for four years, and claim more lives than any other war in American history. It would also be the only war ever fought entirely on American soil, with Americans firing upon and killing Americans.

Members of the New York volunteer fire department formed their own brigades in the United States army. Wanting to stand out from the standard blue uniforms of the Union soldiers, they copied the colorful, Turkish-inspired, uniforms of the French "Zouave" soldiers. They would forever after be known as the New York Fire Zouaves.

Mississippi firefighters countered by forming their own Fire Zouaves units in the army of the Confederate States of America. Unlike their New York counterparts, the Mississippi Fire Zouaves chose as their uniform a bright red wool shirt with long sleeves and wood buttons, and a buttoned-on bib at the front. Their caps were the red "keppie" (from the French Kepi) of the CSA cannoneers, as most of these Mississippi Fire Zouaves operated cannons in the CSA artillery. Their trousers were also wool, either grey, blue, or black, depending on what material was available. The author is fortunate to have in his collection, a modern reproduction of the red keppie cap of a Mississippi Fire Zouave, as worn by those who fired the cannons in the CSA artillery. This red keppie has a flat, black leather front brim. The top of the cap slopes down at the front. On top of the cap is a brass insignia featuring two crossed cannon barrels, and at the cap sides are brass buttons lettered CSA.

Typically, it took a crew of at least three Mississippi Fire Zouaves to fire a cannon. One stuffed the cannon ball into the gun barrel, and ramrodded it in. He then stood behind the cannon, holding his ramrod vertically as a signal that he was ready to have the cannon fired. The second man then lit the cannon fuse, took his place at the back of the cannon, and held a second ramrod vertically as a signal that he, too, was ready for the firing. The sergeant or other officer stood off to the side, then raised his hand as a signal to get ready to fire. When he dropped his hand, the two men behind the cannon pulled a cord that exploded the gunpowder and hurled the cannon ball at Union troops.

The Civil War must have been an odd experience for both New York and Mississippi Fire Zouaves. Their duty as firefighters had been to rescue as many lives and save as much property from harm as possible. Now, as soldiers, their duty was to kill as many of their "enemies," and destroy as much property, as they could.

Americans now remember that at the end of four years, the Union troops defeated the Confederate troops. What is now forgotten is that, for the first two years, the Confederacy was winning most of the battles, inflicting heavy casualties on northern soldiers.

What changed the tide in favor of the Union? A number of factors were involved. Lincoln had appointed, and fired, many generals in charge of the entire Union army. For the first two years of the war, none had proved adequate for the task.

Lincoln had always maintained that the purpose of this war was to preserve the union of all states that made up the United States. It was hard to motivate soldiers to fight for such an abstract ideal. But under pressure from those pushing for the abolition of slavery, Lincoln in 1863 decided to ennoble the conflict by making slavery the issue. He decided that, if he could have but one decisive victory, he would declare the slaves forever free and entitled to full citizenship rights.

Lincoln's decisive victory came on the Fourth of July weekend, at a little crossroads town called Gettysburg, PA, where rag-bedecked southern soldiers had tried to take over a shoe factory to keep themselves equipped with adequate footwear for the long marches they faced daily. This battle over shoes would be the bloodiest three days of fighting in America history.

True to his word, Lincoln followed up the Gettysburg victory by signing the Emancipation Proclamation. This act freed the slaves in the states that were then in rebellion against the United States. Since Lincoln had no authority in states controlled by a government in rebellion, the proclamation did not free a single slave. But it brought a noble purpose to the war, and turned the tide in favor of a string of Union victories to follow.

America's old enemy, England, had just about decided to aid the Confederate cause with money, weapons, and soldiers. But

This oil lamp, on the side of a double-deck, end-stroke hand pumper at New York City Fire Museum, lit the way to many a night fire for 19th century volunteer firefighters. Firefighters also carried torches, which burned coal, wood, or even rags, to light their way at night. Even early steam fire engines sported ornamented whale-oil lamps, with the fire company name etched in the colored glass lenses, before Edison's electric lights made such lanterns and lamps obsolete.

after Lincoln's proclamation, aiding the Confederacy would place the English parliament squarely against human rights and freedoms. So the South would see no aid from England.

Another deciding factor for the Union and against the Confederacy was that the north's economy was largely industrial, with enough factories to keep the Union soldiers equipped with rifles, cannons, ammunition, uniforms, and any other supplies they needed, for as long as the war might last.

The South, on the other hand, had few factories, and was mostly an agricultural economy. Nearly half its people were slaves, and they were not about to take up weapons to help their masters keep them enslaved. Most of the agriculture depended on the slaves to pick the crops, and again, why should they pick crops to feed the soldiers who were fighting to keep them in slavery?

What ultimately won the war for the

Union, however, was General Ulysses S. Grant. He had won several campaigns against the Confederacy on the western frontier of the war, in places like Vicksburg, Mississippi. Placed in command of the entire Union army, Grant quickly grew to understand, as none of his predecessors had understood, the mind of his nemesis, Confederate General Robert E. Lee. This understanding allowed him to encircle the CSA troops and squeeze them into an ever-smaller area. Trapped in the Virginia woods, with no food or supplies, no means of escape, and Union soldiers on all sides of him, Lee finally surrendered to Grant at Appomattox Court House, VA, in 1865.

Now came a period of rebuilding and healing for a weary, war-torn nation. Lincoln and Grant were eager to see the southern states fully accepted back into the Union. American firefighters apparently shared this eagerness. After the war, the red wool shirt with the front bib, the flat-brimmed cap, and black or navy blue wool trousers, as worn by Mississippi firefighters in the Confederate army, became the standard uniform of firefighters in nearly every city and town in America.

Today, the town of Columbia, CA, is restored to how it looked in the post Civil War era, and it is staffed by people dressed in the costumes of that era. Red-shirted volunteer firefighters still pull their hand pumpers and hose reel carts through the town's streets, for the benefit of tourists. And in the town's general store, you can still buy new red woolen firemen's shirts with the front bib and the wooden buttons. I have one of these Columbia firemen's shirts, and wear it with my replica Mississippi Fire Zouaves keppie cap, and original 1870s firemen's leather parade belt, when portraying a firefighter of the Civil War era at various historical reenactments.

Another innovation just after the Civil War was the introduction of the fire alarm telegraph.

In the 1840s, several experiments were conducted with telegraph systems, but none had proven very successful. Samuel F.B. Morse, a renowned portrait painter, learned of these experiments. He strung telegraph wires on trees and poles between Baltimore and Washington, using the broken-off necks of glass bottles as insulators. On May 24,

1844, Morse tapped out in the "Morse code" named for him, the first and most famous telegraph message, to Congress: "What has God wrought?"

In 1845, New York hung bells in various towers around the city, and worked out a system of codes on the bells that indicated the section of the city where a fire was spotted. This bell code summoned the appropriate volunteer fire companies to the correct fire location.

In 1850, Charles Robinson of New York set up a Morse telegraph system between these fire bell towers, speeding up the process of sending fire alarm signals. Someone had to be in the towers to receive the telegraph, and to ring the bells that summoned volunteer firefighters to the firehouse nearest the fire.

In 1852, the same year that Cincinnati introduced the steam fire engine and the paid fire department, Dr. William F. Channing, a 32-year-old physician, installed the first practical fire alarm telegraph system in Boston. The system, funded by a $10,00 appropriation from the City of Boston, consisted of fire alarm telegraph boxes in locations of greatest fire danger around the city. Turning a crank on the box six times, sent a telegraph signal to a central telegraph office, where a telegraph operator retransmitted the signal to the appropriate Boston fire station.

Each fire alarm box also contained a traditional telegraph key, for firefighters to send and receive Morse-coded messages between the box and the central telegraph office. Grounding the electricity inside the box were three brass strips, separated by India rubber cloth, all mounted on varnished wood covered with a glass plate. A local homeowner or shopkeeper was entrusted with the key to the box, as were the neighborhood police officers and firefighters, so that in case of fire, someone could unlock the alarm box and turn the crank that sent the fire alarm signal.

Channing and his business partner, Moses G. Farmer, were granted Patent #17,355 for their fire alarm telegraph system on May 13, 1854. In March, 1855, Channing presented his idea in a speech to telegraph enthusiasts, at the Smithsonian in Washington. In the audience was 33-year-old John Nelson Gamewell, postmaster of the little town of Camden, South Carolina.

The author sports late 19th century fire uniform. The navy blue felt cap, made in San Francisco, has a flat, black-leather brim, and sports a brass 'scrambled eggs' insignia with the word "Chief." The red wool shirt has a front bib, held on with wood buttons. The black trousers are woolen. The white leather parade belt, by Cairns & Brother of New York, sports fire company name "Revere" in raised white leathers on a black leather background. The red leather belt buckle has a raised white number "2" on it. Finally the black leather shoes still shine despite hauling the hand-drawn hose reel behind. *Marc Goldman*

This 1880s fire hydrant, cast at a foundry in San Jose, CA, now sits in the Empire No. 1 fire house at San Jose Historical Museum. This museum consists of several historic homes and commercial buildings, moved from various neighborhoods in San Jose, to Kelley Park. Besides this hydrant, the old firehouse is now home to an early Gamewell fire alarm system, and six hand-drawn fire engines.

Gamewell was particularly impressed that, in the two years that Boston had used Channing's system to transmit all fire alarms, annual fire loss had dropped to less than $1 per citizen.

Gamewell was so enthusiastic about Channing's invention, that he persuaded his friend, wealthy South Carolina jeweler James Dunlap, to finance purchasing the rights to distribute Channing's fire alarm system in the Southern states, and in St. Louis, where Gamewell had relatives. In 1855, Gamewell met with Channing and obtained the distribution rights in the south and west.

Four years later, in 1859, Gamewell bought out all rights to Channing's fire alarm telegraph, for $30,000. That same year, John Brown led his slave revolt at Harper's Ferry, Virginia, and over the next two years, Gamewell would sell systems only to Philadelphia, St. Louis, Baltimore, New Orleans, and Charleston, S.C. During the Civil War (1861-1865), municipal fire alarm telegraph systems were not exactly anyone's

priority, and Gamewell faced near financial ruin, especially when Sherman's march to Atlanta swept right through Gamewell's hometown of Camden, SC. Because he was a southerner, the U.S. Government confiscated Gamewell's patents, further contributing to his financial woes.

Recognizing that the old South was no more, Gamewell and his family moved to Hackensack, NJ, in 1866. There, fortune finally smiled on John N. Gamewell, in the person of John F. Kennard of Boston. Kennard went to Washington, D.C., armed with $20,000 to buy Gamewell's confiscated patents. Inexplicably, the U.S. government sold Kennard those patents for a mere $80. Kennard then teamed up with Gamewell to form Gamewell, Kennard, and Company in 1867. Their headquarters was in New York City, but their factory was at Newton Upper Falls, MA.

In that same year, 1867, Charles T. and J.N. Chester developed a fire alarm box that used a lever, which was faster than a crank in sending the telegraph signal. On August 1, 1869, the Chesters sold a fire alarm telegraph system to New York City, and a few other communities bought from the Chesters. But they could not compete with Gamewell's head start, which was already giving him the lion's share of the municipal fire alarm market in the U.S. While there had been only four fire alarm systems in American in 1862, Gamewell sold 40 between 1867 and 1872. By 1882, there were 62 fire alarm systems, almost all made by Gamewell. With an industrial boom, and the introduction of skyscrapers in the 1880s, Gamewell sold another 237 system between 1882 and 1892. In 1904, there were 764 separate municipal fire alarm systems in America, for a total of 37,739 fire alarm boxes. Early on, the Chesters realized that they could not beat Gamewell, so they joined him, selling him their patent rights in exchange for partnership in the Gamewell firm.

Also at the close of the Civil War, New York finally organized a paid fire department. They had stubbornly clung to their volunteer traditions, even reorganizing the volunteer department in 1855, when other cities were establishing paid departments. Most other large American cities, such as Boston, Philadelphia, and Chicago, switched from volunteer to paid fire department at war's end.

As America grew after the Civil War, so did its industries, and cities sprawled over ever-larger areas. The inner cities grew crowded with emigrants from rural areas, and immigrants escaping poverty in Ireland, Germany, Italy, and many other nations. Conditions in cities like Chicago (1871) and Boston (1872) grew ripe from Great Fires, which caused far more damage than the Great Fires of the colonial era. Had it not been for steamers and fire alarm telegraphs, the damage and loss of life would certainly have been far worse than the staggering losses actually suffered.

In 1871, Chicago stretched 6 miles from north to south, and three miles east to west, with a population of 335,000. America had been joined by rail from coast to coast just two years earlier, with the driving of the Golden Spike at Promontory Point, Utah, in 1869. But already, 13 major American railroads had an important hub at Chicago.

Texas longhorn cattle were herded overland to such Wild West towns as Dodge City, Kansas. There, they were packed in rail cars bound for Chicago, where the huge Stockyards slaughtered and packed them. The packed meats were then shipped by rail again, eastward to butcher shops in New York and Boston, or westward to San Francisco.

Huge ships from eastern cities made their way through the Great Lakes to deliver raw materials to Chicago's factories, and to return manufactured goods from Chicago to east coast markets. Two canals, the Illinois and the Michigan, connected much of the Midwest and south to Chicago for smaller boats.

But the city was not all industry and commerce. It had not yet outgrown its rural origins, and many a small farm still dotted the city. An influx of Irish immigrants to the city had added many barns full of cows and pigs, and small vegetable patches, to make those families self-sufficient for food.

The summer of 1871 had been unusually hot and dry for the sprawling city on Lake Michigan. Only five inches of rain had fallen between July and October, with only a fraction of an inch of rainfall in the entire month of September. The city had thousands of wooden homes and stores. Its sidewalks were made of wooden boards, and its streets paved with heavy wood blocks. And all of that wood was dry from the summer heat and lack of humidity.

In the first week of October, 1871, 27 separate fires started in this dry wood construction. But Chicago's citizens, and even its 185 firefighters, seemed unconcerned. The previous summer, 1870, the fire department had extinguished 600 fires, many started by carelessly placed lanterns in the straw covering the floors of the many barns in the city. So why should this summer be any worse?

Few recalled that the city's 14 steam fire engines had been heavily overtaxed in the summer of 1870, and with good reason. Similarly sized Cincinnati had 19 steamers of 600 gallons per minute or more, all capable of throwing water less than five minutes after lighting the boiler. By contrast, Chicago's 14 steamers mostly had a capacity of 400 to 500 gallons per minute each, and required 10 to 15 minutes to throw water after lighting the boiler.

On Saturday night, October 7, 1871, the city's newly installed Gamewell fire alarm system triggered the Court House bell, summoning about 90 firefighters to a fire in a planing mill on Chicago's west side. By the time this fire was brought under control on Sunday morning, the fire had completely destroyed four city blocks, killed one, and injured several others. Worse, one of the city's 17 steamers (three more than they had the previous summer), a hose wagon, and a large amount of fire hose, had been burned beyond repair.

The fire continued to smolder all day Sunday, and nearly half the city's firemen stayed on all day, drowning the smoldering wood piles to prevent them from rekindling. By Sunday night, most of the city's firefighters were too exhausted to battle another fire.

Patrick O'Leary, an Irish laborer, had bought his small shack of a house on DeKoven Street in 1864, for a mere $500. He and his wife lived in the back, and rented the front to Patrick McLaughlin, a railroad worker, and his family. On Sunday night, October 8, 1871, Pat McLaughlin threw a party to celebrate the recent arrival of his wife's brother from Ireland. The neighbor across the street, Daniel Sullivan, sat on the board sidewalk in front of his house, lis-

FIRE EQUIPMENT

tening as Pat McLaughlin played his fiddle. Legend has it that, at some point in the evening's festivities, the McLaughlins ran out of milk for the mixed drinks they were serving, and awakened their landlady, Kate O'Leary, to ask if she had any milk to spare. Annoyed, she trundled out to the barn with her pail, lantern, and stool, and half-awake, began milking her cow. The cow, even more annoyed at being milked late at night, then kicked over the lantern, touching off what was, until then, the most destructive fire in American history. This part of the tale is probably false.

At about 8:45 P.M., Daniel Sullivan, from his position across the street, noticed a flicker of flame from the O'Leary barn, where two cows and a horse lived. Sullivan ran toward the barn, shouting 'Fire! Fire!' as he ran. As he passed the O'Leary house, he noticed that that there were no candles or other lights in the O'Leary home at the time. So the O'Learys were evidently asleep, not milking a cow, when the fire started. A lantern had evidently been left unattended in the barn, and a cow had indeed kicked it over, not in anger over being milked late at night, but just in the normal course of its moving around in the barn.

Sullivan managed to save one of the two cows from the burning O'Leary barn. Mrs. O'Leary was not inside the barn when he arrived, she did not pass Sullivan on his way to the barn, and he saw no evidence that she had hastily retreated from the barn when it ignited. Nevertheless, contemporary cartoonists would portray Mrs. O'Leary as an ugly and evil witch, who had somehow conspired with the cow to burn down the City of Chicago.

A tragic flaw in the early version of the Gamewell fire alarm system now became evident. The crank that transmitted the fire alarm was inside the locked alarm box, and only a local shopkeeper had the key. At 9:00 P.M. on a Sunday night, the shopkeeper was, of course, not at his store, and was nowhere to be found. And so, nobody sent a fire alarm. The fire quickly spread through home after home in this working-class neighborhood. Curiously, the flames spread in one direction from the barn, away from the O'Leary house, and while most of the city burned, when it was all over, the O'Leary home still stood!

In the tall tower atop Chicago's Court House, Matthias Schaffer was paid to watch the night sky for fires. At 9:30 P.M., he noticed the flames from the fire that had by now consumed an entire city block, and was still spreading. In the dark, though his telescope over a distance of several miles, it was hard to judge the precise location of the fire, and Schaffer guessed that the nearest alarm box was Box 342. He instructed young fire alarm telegraph operator William J. Brown to send the signal for alarm box number 342. But Shaffer had guessed wrong, and this alarm sent fire companies from a different neighborhood, to an alarm box where there was no fire.

Those firemen spotted the flames from the DeKoven Street fire, and seeing that here was no fire anywhere near Box 342, they took it upon themselves to go to where the flames were. But coming from the wrong neighborhood, and going to the wrong box first, further added to the delay in getting water on the fire, which had already been raging for 45 minutes before the first alarm even sounded.

Meanwhile Shaffer, peering more intently into the night sky, realized his earlier error, and told Brown to transmit the alarm to the correct alarm box this time. Perhaps out of youthful pride, Brown refused to acknowledge that the first alarm he sent was a mistake, and stubbornly transmitted the alarm as Box 342 again. This sent more firemen and fire apparatus to the wrong location, giving the fire still more time to rage destructively out of control. In those days before two-way radios, and five years before telephones were invented, there was no quick and easy way for the first fire companies to let Shaffer, Brown, or the second-alarm fire companies, know that Box 342 was not the fire location.

Meanwhile, Joseph Lagger, stoker on the Amoskeag steamer of Little Giant Engine Company Number 6, spotted the DeKoven Street fire. The Fire Company's foreman, Henry Musham, took it upon himself to take his men and engine to the fire, without waiting for the fire alarm telegraph office to dispatch them there.

When the first alarm sounded, Matthias Benner was at his home on Randolph street, resting from directing operations at the previous night's fire. Benner was Fire Marshall

for the entire West Division, one of the city's three fire districts. Dressing quickly, he scurried to the nearest street corner, Randolph and Jefferson, where the fireman assigned as his driver picked him up. Benner immediately realized that the alarm for Box 342 was an error, and directed his driver toward the flames shooting skyward from the O'Leary barn. Musham and the crew of Little Giant Engine 6 were already at work on the fire when Benner arrived.

Benner noticed that some stores and saloons in the next block were burning. He directed Musham to recruit the citizens watching the fire, to help set up a second hose line from the steamer to this second block. A few spectators did help for a while, but as the roaring fire advanced toward them, they dropped the hoses and ran. A chance to stop the fire when only two city blocks were burning, was thus irrevocably lost.

By 11:00pm, a strong wind started to blow, spreading the fire northward. About 1:30am Monday, October 9, the fire had spread to engulf the Court House several miles away from the fire's origin, from which the first erroneous fire alarm for Box 342 had been sent. It then quickly marched north, crossing the river near the State Street bridge and spread all the way to the Water Tower, cutting off the flow of city water on which the fire engine depended for their streams.

Before dawn, most of the South Side of Chicago was in flames, and panicked residents fled over the Randolph Street Bridge to what they thought was safety in the West Division. Hotels and grain elevators were burning directly behind them, and the flames were headed their way.

About 7:00am on Monday, the flames set fire to the rigging of ships on the Chicago River. This gave the fire, which had already destroyed most of the West Side, a chance to leap the supposed boundary of waterways and spread to other sections of the city.

By 1:00pm Monday afternoon, an area six to seven miles long by a mile wide was already in ashes, as firemen and fire engines arrived from all over Illinois, and from as far away as Milwaukee, St. Louis, Cincinnati, and Dayton.

Ahrens Mfg. Co. sent one of their new steamers, not yet delivered, to help fight the fire, and company president Chris Ahrens accompanied the engine to Chicago, to operate it there himself. He stuck with the engine for three days. This gesture of good will would be amply rewarded. Up to then, Chicago had never bought an Ahrens steamer, but between 1873 and 1912, the city bought 120 steamers from Ahrens, and when Ahrens switched to making motorized fire engines, Chicago would buy 57 pumpers of this make between 1916 and 1929.

At 11:00pm Monday, rain began to fall, and by late Tuesday morning, October 10, 1871, it was all over. All that was left to burn was the prairie grass north of the city, where the fire mercifully stopped. The last house to burn, on Fullerton St., caught fire about 3:00am. When it was all over, 3.3 square miles of Chicago lay in ruins, with 17,000 buildings destroyed. Insurance loss was pegged at $200 million of those far more valuable 1871 dollars, including $88 million in actual property damage. 90,000 people were left homeless, 250 died in the fire, and another 50 died soon after from burns or other injuries suffered in the fire. It took another full day to drown all of the smoldering embers, so that the fire could not rekindle. Firefighters from all over the Midwest did not pack up and head home until Wednesday, October 11, 1871.

Just over a year later, Boston would suffer a similar fate. The city, already two centuries old, was full of crooked streets crowded with wood and brick business blocks stacked against each other.

On Saturday night, November 9, 1872, a fire started in the rear of a hoop-skirt factory near the top of a six-story industrial building at Summer and Kingston Streets. Because downtown was deserted for the weekend, nobody noticed the fire until flames shot from all six floors of the building. Someone started screaming "Fire! Fire!" and Patrolman John Page of the Boston police department heard the cry from several blocks away. He came running, but did not have a key to Fire Alarm Box 52 on that corner (today, Boston's fire buffs club is called Box 52 in remembrance of the Great Fire of 1872). So Page had to run to Beech Street, get the alarm box key from the Patrolman there, run back, open the box, and turn in the alarm. By 7:08pm, the fire was already big enough that several people spotted it miles away in Charlestown, yet the first alarm was not sent until 7:24pm.

When Fire Chief John S. Damrell arrived at the fire about eight minutes later, the entire building was a wall of flames, and the heat was so intense that he could not stand within a block of it, much less send firemen inside with hoses. Overheated pieces of the granite walls, weighing from one to 20lb each, were flying everywhere, setting new fires wherever they landed. Damrell immediately called for a General Alarm. This meant that every fire engine in the city was to respond immediately to this fire. All city firemen, whether on duty or off, were also required to respond.

But the Boston Fire Department was under an unusual handicap. About two weeks earlier, a disease called 'epizootic' had struck nearly all of the city's fire horses, making their feet and legs cold and swollen. While only four of the department's horses died, most were still too weak to stand upright in their firehouse stalls, much less haul a heavy steamer, ladder truck, or hose wagon behind them through the city streets. Only five days before the fire, 70 or 80 of the department's 95 horses had been in animal hospitals, under constant veterinary care, before returning to their fire stations.

The night of the fire, only one of the city's steamers was brought to the blaze by its regular team of horses. Four engine companies borrowed horses from various stables, but none of these horses were strong enough and well trained enough for the task, and they showed signs of great strain when they arrived at the blaze.

Boston's remaining 16 steamers, all weighing three tons or more, had to be hand-pulled to the fire by long lines of firemen, as in the hand pumper era a century earlier. Firemen also hand pulled all of the department's hose carriages and ladder trucks to this fire.

Most of the buildings in this neighborhood had French style Mansard roofs. These were false roofs sticking above the actual roof, with plenty of space for fire to hide undetected between the inner and outer roofs. As the granite walls of building after building crashed to the street, the heavy stones ruptured gas mains, feeding the hungry flames still further. Worse, these falling walls crashed on top of fire hydrants, ren-

Yet another of the fancy 19th century parade hose carriages on four wheels that seemed so popular with volunteer firefighters, even though they had almost no practical firefighting use. This one is displayed at New York City Fire Museum.

During the American Civil War, volunteer firefighters from New York City organized their own regiment. Their colorful uniforms, far from standard Union Army issue, were modeled after French Zouave soldiers and were dubbed the Fire Zouaves. During one battle, they capture this "Philidelphia-style" hand pumper, the *Hope*, from the South. It's now displayed at New York City Fire Museum.

At 3:00am November 10, 1872, 42 steamers and 1,000 firefighters made a stand against the fire, along Washington and Milk streets. Despite the terrific heat blasting toward them, and the falling walls showering hot granite upon them, the firemen stood their ground, aiming 100 streams directly into the advancing wall of flames. The Old South Church on that corner crackled, hissed, and steamed, but the barrage of water kept it from igniting. The fire did not spread any further north or west of that intersection, and after the fire had burned more than 50 city blocks, the long night's battle was finally over.

While the Great Chicago Fire had ended with nature's deluge of rain, the Great Boston Fire ended with a man-made deluge of 100 fire streams.

In that fire 13 firemen died, nine of them from other fire departments, on mutual aid call, making the ultimate sacrifice to help Bostonians on their night of greatest need. It is nights like November 9, 1872, that remind us of the proud traditions of America's fire service, and how much we all depend not only on their knowledge, skill, and bravery, but also the effectiveness of their equipment. At the start of this blaze, the equipment on which Boston's firefighters relied, failed them: no alarm-box key was available, and the horses that pulled their apparatus were sick. But their equipment, in the form of 42 steamers, 100 hose lines, and 100 brass nozzles, ultimately saved the day.

The back-to-back Great Fires taught America's firefighters, and its fire equipment dealers, many valuable lessons.

For example, John Gamewell learned to make his fire alarm boxes easier for the general public to operate. He added a hook to the outside of his fire alarm boxes; pulling down the external hook tripped the internal lever, and any citizen could now send the fire alarm without needing a key to open the alarm box. Only firefighters would now have keys, to open the box after a fire and reset the working mechanism, so it could transmit an alarm signal for the next fire.

Eventually, nearly every fire alarm box in America would use the identical standardized Gamewell key. Firefighters would not have to fumble to figure out what key fit which box, and Gamewell did not need to keep track of which key fit which city's

dering them useless and forcing firemen to flee, often abandoning hose and apparatus in the fire's path.

While in the Great Chicago Fire the previous year, a strong wind fanned the flames, the wind at the Great Boston Fire of 1872 was mild. As a result, the Boston fire did a curious and unexpected thing: although what little wind there was blew from the north toward the south, the fire spread AGAINST the wind, destroying buildings ever further to the north.

During the night, 45 engines, 52 hose wagons, 3 ladder trucks, and 1,689 firemen arrived from out of town to aid the Boston firefighters. Every town within a 50-mile radius of Boston sent help, as did cities further away, from Biddeford and Portland, Maine, to Norwich and New Haven, CT.

boxes when someone ordered replacement keys. I have several fire alarm boxes and several Gamewell keys in my collection. I once found an old alarm box for sale in an Antiques store, but it lacked a key, so I used one of my keys to open the box and inspect the condition of the working mechanism inside. The amazed shopkeeper asked me how I happened to have a key that fit her alarm box!

The 1880s would be the most written about and romanticized era in American history: the cowboy era. The generals and soldiers who had fought in the Civil War, now turned their attention to taming the Wild West, and exterminating the Native American "Indians" who stood in the way of Manifest Destiny, the idea that America had a right to conquer the continent from coast to coast. The most successful Civil War generals now gained even more fame as Indian Fighters: Phil Sheridan, William T. Sherman, and of course, George A. Custer of Little Big Horn fame. As the Indians were pushed ever further west, cowboys moved in first, then farmers, and finally the businessmen, industrialists, and railroad tycoons who would build cities and towns ever farther to the west towards the Pacific.

With each new settlement, came the danger of fire, and the need for fire departments. The shrewder of the dealers followed this westward migration, selling their fire engines and firefighting equipment to cities and towns ever further to the west.

Many of the settlers in these new towns had fled revolutionary war in Germany, or grinding poverty and famine in Ireland. Both Germans and Irishmen were hard working and hard drinking, and an ethnic rivalry between the two immigrant groups existed nearly everywhere that new settlements were built.

In the 180s, two of the most successful builders of steam fire engines, and marketers of firefighting tools and equipment, were German-born Chris Ahrens of Cincinnati, and Horace Silsby of Seneca Falls, NY, son of Irish immigrants. A look at sales records of both companies in the 1870s and 1880s, shows that both men followed the westward migration, selling fire engines an equipment in Illinois and Missouri, later in Kansas and Nebraska, still later in Colorado and Utah, and by the 1890s, in California and Oregon.

But the records also show another interesting pattern. Ahrens's sales tended to be in towns with Germanic names, such as Frankfort, Leipsic, or Germantown, and the mayors and fire chiefs had names like Graf, Frantz, Bache, Fleishmann, or Goetz.

Silsby sales, on the other hand, were in towns with names like Connemaugh, Danielson, Mahanoy, Shamokin, or Sharon, or with more traditional English names such as Avon, Bradford, Cambridge, Chelsea, Norwich, Oxford, Portsmouth, Reading, and Salisbury. The mayors and chiefs, who chose to buy Silsby engines for their towns, had names like O'Hara, Sullivan, or Shaughnessy.

In communities with mixed German and Irish populations, Horace Silsby and Chris Ahrens, and their steamers, often went head to head for a sale. An amusing example of this took place in Watertown, WI, in 1876. Watertown was aptly named: a river ran through the center of town, and the river delineated the living areas with almost all the German population on one side of the river, and the Irish on the other side.

All sorts of competitive tests were tried on the fire engines, with Silsby winning one competition and Ahrens another, for several days. Naturally, the Irish population wanted a Silsby, and the Germans wanted an Ahrens. City officials, who wanted both camps to re-elect them, could not decide which engine to buy. The Gordian knot was cut by buying both engines, building two firehouses, and organizing two fire companies. The Ahrens was quartered on the German side of the river, and the Silsby on the Irish side.

Along with westward expansion in the 1880s, came upward expansion. Older cities with growing populations had no room to expand outward, so they started building upward, erecting ever-taller buildings. Fires in the upper floors of these "skyscrapers" were beyond the reach of conventional ladders and fire hoses. In the 1880s, America's inventive genius met this challenge with three new weapons in the firefighter's arsenal: the aerial ladder truck, the water tower, and the standpipe. We shall examine each of these, and their ancillary equipment, in the next chapter.

CHAPTER FOUR

Fire Equipment in the Steam Era 1881-1909

In 1866, Amoskeag of Manchester, NH, sold one of their steam fire engines to San Francisco. One of their salesmen, Daniel Hayes, accompanied the engine, to deliver it and instruct the firefighters in its operation.

It would be three years before the first transcontinental railroad opened, and 90 years until the Interstate Highway system. Even the Panama Canal was decades into the future. So Hayes and the steamer had to travel south through the Atlantic, around Cape Horn at the southern tip of South America, and north up the Pacific coast to S.F. It was a very long journey indeed. So Hayes appreciated the warm reception which San Francisco firefighters gave his arrival, and he was easily persuaded to settle in that city. His mechanical knowledge of fire apparatus quickly earned Hayes the post of Master Mechanic for the San Francisco fire department.

In 1867, Hayes began work on a concept that had been on his mind for some time. The longest fire department ladders of that day extended to 65 or 75ft, and were extremely heavy, requiring a large number of men to remove them from the ladder racks, and to carry, position, and hand raise them. In this industrial era, and with a booming post Civil War economy, there had to be a way to mechanize the process of raising long ladders, and to sell this mechanized equipment in enough quantity to be profitable. Hayes proposed attaching the tallest ladder to the base of the ladder truck, and using mechanical means to raise, lower, and extend it. Assisted by several San Francisco firefighters, Hayes completed his prototype aerial ladder truck, the world's first, in 1870.

After the Great Chicago Fire of 1871 and the Great Boston Fire of 1872, the need for the Hayes aerial ladder became obvious, and he sold them from coast to coast. His success encouraged a host of imitators trying to cash in on this new market. But many of these cut corners on materials and workmanship to make a quick profit, giving the aerial ladder a bad name.

Italian inventor Paolo Porto was one of the first to copy the Hayes aerial, making his ladders of the finest Italian white walnut. He persuaded Mrs. Marie Belle Scott-Uda to sell these aerial ladders in the U.S. for him. She brought three of Porto's aerial ladder trucks to America in 1874, and the demonstration she staged at New York's City Hall Park so impressed city officials that they bought all three for $25,000.

William B. White, secretary of New York's board of fire commissioners, held up paying Mrs. Scott-Uda, however. He wanted her to sell him the three ladder trucks, plus the U.S. patent rights to them, not for the agreed on $25,000, but a mere $16,000.

White had four more aerials built on the Porto patent, at a factory in Concord, NH. Far from using Italian white walnut, White acquired the lowest priced, lowest grade wood he could find, to increase his profit.

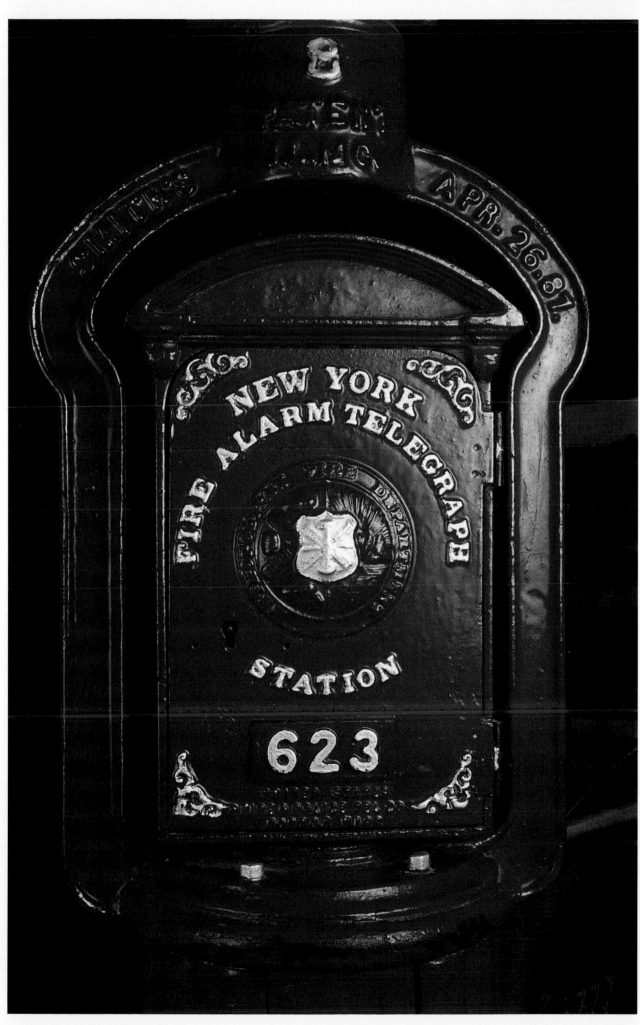

Fire Alarm Box number 623 of New York City bears a patent date of April 26, 1887. It is now displayed at New York City Fire Museum.

White demonstrated one of his "Scott-Uda" aerials, as he was calling them, at New York's Rutgers Square. Seven firefighters climbed the ladder, spaced 10 to 12 feet apart. When three of them were on the fly (extension) ladder, the wood began to twist to one side, finally snapping in several places. The three firemen on the fly ladder were hurled to their death, and the four on the bed (main) ladder were severely injured. Mrs. Scott-Uda, under whose name these aerials were marketed, had been watching the demonstration, and fainted.

An inquiry afterward condemned White for shoddy workmanship, and the board of fire commissioners for not scientifically testing the aerial before permitting seven firefighters to climb it. The New York fire department was prohibited from any further use of not only White's four low-quality aerials, but even the three higher-quality Porto aerials they already had.

The birth of skyscrapers, office buildings more than 10 stories high, in the 1880s made the need for aerial ladders obvious. Hayes had never compromised quality on his aerials, and in the 1880s, they regained and surpassed the popularity they had enjoyed in the early 1870s.

By this time, Hayes had done away with the laborious and back-breaking hand cranking required to raise his early aerials. Instead, he compressed two massive springs, one on each side of the ladder, inside a cylinder. Releasing the spring tension with a lever, easily and rapidly raised the bed ladder, but firemen still had to hand crank to raise and lower the fly ladder. When the fire as out, the lever compressed the spring back into its cylinder, safely and easily lowering the aerial ladder back onto the truck.

Hayes eventually sold his patents to American-LaFrance of Elmira, NY, and they used the spring raise on all of their aerial ladder trucks through the 1930s.

In the early 1900s, fire chief Edward F. Dahill of New Bedford, MA, developed a pneumatic aerial hoist. Instead of a spring, the lifting cylinders contained compressed air. It was easier to compress and decompress air than it was a steel spring, and air did not wear out and lose its tension as springs could. Air raised the ladder more rapidly and smoothly, with less jerking, than a spring did.

At least one careless mechanic once loosened the cylinder cap of a Hayes-style, spring-raised aerial ladder to investigate why the spring was not decompressing and raising the ladder. The spring did decompress, knocking him in his chest, and sending him flying head first through a brick wall to his death. With the Dahill air hoist, if you were so inclined to uncap and examine the cylinder, the worst that could happen was you would face was a blast of escaping air. Dahill, while still serving as New Bedford's fire chief, started a company to build and sell his aerial ladders. He sold a dozen or more of these aerials to cities all over the U.S. between 1904 and 1912.

Dahill eventually sold his patent rights for the air hoist to his lifelong friend, Charles H. Fox of Cincinnati. Fox, a former Assistant Chief of the Cincinnati Fire Department, was son-in-law of steam fire engine builder Chris Ahrens, and later partnered with several of his in-laws as president of Ahrens-Fox Fire Engine Co. Ahrens-Fox built about 70 aerial ladder trucks with the Dahill pneumatic hoist between 1916 and 1937. They tended to raise more smoothly and rapidly than the spring hoists.

Another method of raising aerial ladders that found some popularity in the 1890s and 1900s was hydraulics. Instead of a spring or compressed air, the lifting cylinders contained water under pressure. Releasing the water pressure raised the aerial, and compressing it again lowered the ladder. But it was hard to maintain a water-tight seal, and if water leaked out while the ladder was raised and firefighters were on it, the ladder and its occupants would come down very quickly. Water evaporated, so it needed replenishing between uses. The air hoist had its own compressors to keep the lifting cylinders filled with compressed air when not in use.

In the 1930s, a new type of hydraulic aerial, using specially formulated hydraulic oils piped through hoses, made the hydraulic aerial hoist superior to every other method. Hydraulics raise all aerial ladders made today, and the spring and air hoists are now found only in the hands of collectors and museums.

In the late 1930s, one other method of raising aerial ladders was briefly tried with much fanfare and unfulfilled promise: electricity. A series of heavy-duty electric batteries powered a set of electric motors, and with a maze of wiring, the electricity raised and lowered the aerial. Ahrens-Fox Company of Cincinnati built three aerial ladder trucks with electric hoists in 1939, and amazingly, all three survive today, although all three have been abused and neglected. The need for 48 gigantic truck batteries, two massive electric motors, and miles of cloth-covered copper wires (any one of which might short-out at any time), made these aerials far less practical than the new breed of hydraulic raises. Today, electric aerial ladders are a mere historical curiosity, a minor footnote in American fire service history.

If it carried nothing else but a long, easily raised and lowered ladder, the aerial ladder truck would have been a marvelous aid to the firefighter. But by 1900, the typical aerial ladder truck also carried: 14 or 15 hand ladders of various lengths (ranging from 10 to 35 feet long); lanterns, axes, plaster hooks, and ropes; fire extinguishers; extra hose, nozzles, couplings, hose repair kits, and buckets; tools for prying and ramming doors open; saws, mauls, picks, hammers, and chisels; roof and wire cutters; acetylene torches; a folding life net, rope gun, smoke masks, and rubber gloves; a first aid kit; and searchlights.

On other manufacturer would make its reputation with ladder trucks, and eventually branch out into many other types of fire apparatus: Seagrave of Columbus, OH.

In the late 1870s, 32-year-old Frederic S. Seagrave began making wooden ladders in a shed behind his home in rural Rochester, Michigan He sold them to local farmers to use in picking fruit on their orchards.

Born June 16, 1849, Seagrave grew up on a farm in Massachusetts, but had more of a taste for inventing and adventure than for farming, so had moved West as a young man. At age 20, he married Adelaide Rutledge of Detroit, and became a salesman, tinkering with new inventions in his spare time.

In 1881, Seagrave's tinkering led to the grant of his first patent, for a trussed wood ladder. Seagrave had strung a heavy wire from corner to corner along both sides of his ladder, and pulled it taut, giving the ladder extra strength and rigidity. Local volunteer firemen found Seagrave's ladders to be

Elkhart Brass used an elk as their emblem for many years. This Elkhart copper soda-and-acid fire extinguisher is on the running board of an 1890s horse-drawn ladder truck at New York City Fire Museum.

In the 1890s, the driver of this horse-drawn hose wagon used a foot pedal to ring the front gong, clearing New York's busy streets on the way to a fire. The hose wagon is now in the New York City Fire Museum.

Knurled edges were common on nozzle tips that fit on the playpipes of a horse-drawn steamer or hose wagon. These rigs usually carried two playpipes, and a range of different diameter nozzle tips to fit on them. This one, in the author's collection, dates from about 1900.

even stronger, and provided a hand railing for firefighters to hold while climbing the ladder. In an 1891 patent, he added another feature: steel bands were heated, wrapped around the truss and the ladder at regular intervals. When these bands cooled, they contracted to hold the truss tightly to the ladder, for even greater rigidity. Seagrave, now a division of FWD, still uses this same basic method of making trussed ladders for its ladder trucks today, over a century later.

A typical Seagrave hand-drawn or horse-drawn city service or village ladder truck, built in the Detroit factory, carried a 40ft extension, 14ft and 16ft roof, 12ft scaling, and 18ft wall ladder. It also carried a pike pole (to pull down walls and roofs), a ladder lifting pole, four to eight rubber fire buckets, an axe, a torch to light the way at night, two Dietz hand lanterns, a front gong, one or two fire axes, heavy-duty rope, and chains.

In the mid to late 1880s, those who had made their fortunes in the Michigan lumber business, sought ways to invest their riches in new industries. Many started steel mills, and the character of Detroit changed dramatically. The city became dirtier, noisier, and more congested. Capital for small businesses like Seagrave's ladder operation dried up, as it was all invested in the large steel plants. By 1900, these lumber barons found a new industry to invest in: automobiles. While the center of the fledgling auto business had been Ohio and Indiana, it shifted to Detroit by 1910, as automobile makers (such as Packard of Warren, Ohio) moved to where the investors were.

With the lumber barons moving into new industries, not only the supply of capital, but also the supply of the inexpensive, high-grade lumber that Seagrave needed for his ladders, began to shrink. The skilled craftsmen he needed were moving into new industries, and the extra buildings he could not compete with huge steel makers for real estate. Land had become too valuable for him to keep renting buildings for painting and drying fire engines.

Seagrave found a much better climate for his fire ladder business farther south, in Columbus, Ohio. He began moving his operation there in 1891, and in 1893, he opened his new Seagrave & Company plant on Lane Avenue in Columbus. This plant

strong and sturdy enough to support several men, carrying their heavy tools and equipment up the ladder with them. The wire trusses kept the ladders from bending, flexing, or snapping under the great weight. Soon, Seagrave received more orders for ladder from fire departments than from fruit orchards.

Inevitably, firefighters began asking Seagrave to make the horse-drawn wagons to carry his ladders to fires. He could not do this in his little shed at Rochester, Michigan, so in 1881, he founded Seagrave & Company, with a small factory at 418 Michigan Avenue in Detroit. The upper Michigan peninsula was a prime source for the high-grade lumber Seagrave needed to make his ladders, and the lumber barons had capital to invest in businesses like his. Soon, he was making ladder trucks, two-wheel hose reels, and four-wheel hose wagons. He hired skilled craftsmen to paint, stripe, and letter his fire engines, and began renting various buildings around Detroit to paint and dry his vehicles (paint and varnish of that day took up to a month to dry).

In 1883, Frederic Seagrave patented a better way to truss his ladders. Instead of stretching wire taut from corner to corner, he steamed and bent a long, thick piece of hardwood as a truss. This made his ladders had the advantage of being on a railroad siding, where raw materials could be shipped in, and completed fire engines shipped out. Skilled craftsmen were plentiful, and the lower humidity allowed paints and varnishes to dry more quickly than in Detroit.

In 1898, Julius Stone, a Columbus banker and owner of Ohio Buggy Company, became a major investor in Seagrave's thriving fire apparatus business. The company name then changed from Seagrave & Co. to The Seagrave Company.

Four years later, in 1902, the volume of fire apparatus sales had outgrown the plant's capacity. Frederic Seagrave was now 53, and decided that he did not wish to expand his business again. So he sold out to Julius Stone, who promptly moved Seagrave into his Ohio Buggy Company factory, at 2000 High Street, Columbus, Ohio. Seagrave fire engines would be made in that plant until 1965, when truck manufacturing FWD Corporation bought Seagrave out, and moved operations to their plant in Clintonville, WI. Seagrave fire engines are still built in Clintonville today.

But Frederic Seagrave was not yet out of the fire engine business. He retained the right to build and sell fire engines in Canada, and set up a factory in Walkerville, Ontario, across the river from Detroit, where his business had begun 20 years earlier. He placed his son, Warren Edmund Seagrave, in charge of the new W.E. Seagrave fire apparatus company. This new company distributed fire engines in Canada for the Seagrave plat in Columbus, and later built fire engines there to the same designs that the Columbus plant used.

Freed from the responsibility of running the Columbus factory, in 1902, Frederic Seagrave concentrated on a new invention: a spring-raised aerial ladder that was superior to those of Daniel Hayes. He completed the prototype in 1903, and had perfected it well enough by 1905 to sell and deliver several of them that year.

Seagrave's aerials pivoted at a lower point on the ladder than the Hayes, for a lower center of gravity. The two main lifting springs, near the base of the ladder, also counterbalanced the main ladder's weight. This made raising and lowering the aerial faster than the Hayes design, and made it more stable when raised. Like Seagrave's

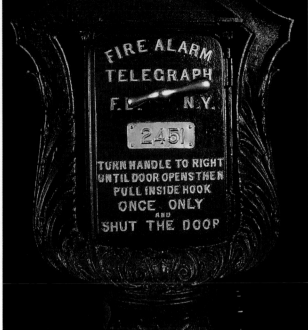

Left: The "keyless" fire alarm box door, where a simple turn of the handle opened the box, appeared in 1875. Before that, a local police officer or shopkeeper was entrusted with the key, resulting in tragic delays while a key was found before an alarm was sent. Boxes with curved tops, as an alternative to the more familiar peaked-top style, appeared in 1890. Many of these 19th century fire alarm boxes, such as #2451 of New York City, saw more than half a century of continuous use. This one is now on display at New York City Fire Museum.

Left: This fancy 1905 ribbon, now at New York City Fire Museum, identified the wearer as a delegate to the 33rd annual New York State Firemen's convention.

ground ladders, his wood aerial ladders were trussed for greater rigidity.

Two hand crank wheels on either side of the Seagrave aerial's base, connected to a long screw shaft and bevel gears, extended the fly ladder.

Aerial ladder trucks were soon common in American cities. And they carried just about everything to a fire, as this impressive list for a New York City ladder company of 1911 shows: Axe; life belt; ladder strap; large and small maul, rivet cutter, cotton hook; 2 signal flags; shovel, wire cutter, Hale door forcer, hay fork, ram; Milburn light, 2 red lanterns; gas wrench, gun, claw tool, life net, megaphone, medical bag; junior searchlight; 5ft, 10ft, 12ft, 15ft, 20ft, and 25ft hooks; body bag; Vagen-Bader gas mask; Pyrene fire extinguisher; cellar and sub-cellar pipes.

The aerial ladder, such as those built by Hayes, Dahill, and Seagrave, was a great way for firefighters to climb up to higher floors of burning buildings, carrying a variety of hand tools with them. They could also rescue citizens trapped in a fire, and carry them down the aerial ladder. But the ladders themselves were made of wood, and were not strong enough to support the weight and back-pressure of a fully-charged fire hose strung to its top. To deliver large volumes of water at great pressure to upper stories of burning buildings, the wooden aerial ladder was not the solution.

Fortunately, there was another solution for delivering water to fires in tall buildings: the water tower. A water tower consisted of a steel pipe on a wagon, which could be raised to a vertical position. One or more pumpers pumped water into the base of this pipe. A nozzle atop the pipe sprayed large volumes of water, at great pressure, into the upper floors of burning buildings. Water towers ranged in height from 35ft (able to spray water into a fourth-floor window) to 76ft (delivering a stream to the 9th or 10th floor) when raised.

The first patent for a water tower was granted to Lester Day of Buffalo, N.Y., on October 27, 1868. He apparently never built his fire engine, and if he had, its spindly design would probably not have withstood the tremendous water pressure pumped through this type of machine.

The first water tower, called a "hose elevator," elevated a fire hose, placed in a bucket-like platform, to a working height of about 40ft (the base of a fifth-floor window). Mr. Skinner of Chicago invented and built it in 1869, and Chicago ire Department placed it in service as Hose Elevator Number 1. It saw service at the Great Chicago Fire of October 9-11, 1871. In 1873, Chicago Fire Department removed its hose platform, so while its mechanism could still be raised and lowered, it seemed to have no further practical use. By 1881, it had been scrapped.

The following year, 1870, a Mr. Knockes of Chicago developed a wagon with a platform that could be raised and lowered to paint the exterior of multi-story buildings. George W. Hannis of Chicago adapted one of these Knockes painter's platforms into a hose elevator, which was placed in service that year as Hose Elevator Number 2 in the Chicago Fir Department. Like the Skinner hose elevator, the Knockes-Hannis performed well at the Great Chicago Fire in 1871. About 1876, the wooden hose platform was damaged by the thrust of the hose under pressure, and by 1881, it had also been scrapped.

On April 22, 1876, John B. Logan of Baltimore, MD, applied for a patent on a water tower, which was granted November 21 of that year. It was to consist of a 12ft pipe attached to a wagon bed, and two more 12ft pipes that could be attached to he main pipe. Firemen wold then hand crank the entire 36ft of pipe to a vertical position. But Logan lacked the financial backing to build it.

Two years later, in 1878, Logan made two improvements to his water tower design. H increased the diameter of the wheels on which the wagon was to be mounted, and

he added jacks that could be placed under the wagon, to stabilize it against the great water pressure. With these improvements, Logan convinced local industrialist Abner Greenleaf to finance his invention. Greenleaf and Logan completed their first water tower in 1879, and were awarded a patent for it on April 27, 1880.

On June 14, 1879, Greenleaf and Logan demonstrated their prototype water tower in New York City. On July 2, 1879, the New York Fire Department placed it in service on a trial basis as Water Tower 1, at quarters of Engine 7 on Duane Street in lower Manhattan.

When its three pipe sections were assembled, and the tower cranked to vertical on a massive quarter-gear, its nozzle tip was 50ft above the ground, high enough to shoot water into a fifth-floor window. A series of cables kept its three pipes connected and stable. Another cable, attached to the nozzle at top and wound on a drum, allowed firefighters on the ground to adjust the angle of the nozzle by hand cranking.

On March 30, 1881, after a trial of almost two years, the Fire Department of New York (F.D.N.Y.) finally purchased this water tower from Greenleaf and Logan for $4,000. On December 28, 1882, F.D.N.Y. mechanics added counterweights to the body, and chains to secure the rear axle, so the tower would not overturn while operating at a fire. With these modifications, the first water

Although built 3,000 miles away in Seneca Falls, NY, Rumsey hand pumpers were popular in northern California around 1900. This fione example belongs to the fire department in Redwood City, CA.
Ed Hass

tower served another ten years, before it was scrapped in 1892.

Greenleaf and Logan completed a second water tower in 1881, and delivered it to F.D.N.Y. in November, 1882. Their third tower went in service in Boston on March 20, 1882. The fourth (and last) Greenleaf-Logan tower was built in 1884, and served F.D.N.Y. from 1885 to 1900. It then served in Trenton, NJ, until junked sometime in the 1930s.

In 1886, George C. Hale, Fire Chief of Kansas City, MO, also developed a 50ft, hand-raised water tower. His prototype was briefly used in Kansas City. In 1888, he built a second hand-raised water tower, this one reaching 75ft up, and sold it to Milwaukee. They tested and rejected it, and Hale took it back to Kansas City, which also used it for a time.

In 1888, Hale developed a better way to raise a water tower. The soda-and-acid chemical engine was just gaining favor in America's fire departments. This type of engine mixed bicarbonate of soda with sulfuric acid in a copper tank. The resulting chemical reaction created a CO_2 gas, which forced water out of a small hose onto a fire. This let firefighters extinguish small fires, as in dried grass or trash cans, without requiring a pump. Hale decided to use this same CO_2 gas to push a lifting cylinder and raise the water tower. Slowly releasing the gas would lower the tower. Hale was granted a patent on his chemically raised water tower on May 20, 1890.

Kansas City bought the first of these chemically raised water towers in 1889. It reached to a height of 45ft. Later that year, Chief Hale built a 55ft version, and sold it to Buffalo, NY, where Lester Day had invented the first water tower 20 years earlier. A 65ft version followed in 1895.

Hale later modified his design, so that some of the water pumped from steamers into the tower, was diverted into the lifting cylinders. This did away with the need for the chemical tank, the explosive chemical reaction, and the requirement that firefighters handle sulfuric acid which, if spilled, could eat through their skin. By 1930, 41 Hale water towers had been built.

In 1898, Henry Gorter, master mechanic for the San Francisco Fire Department, modified Hale's design, using a built-in water motor to supply the water that raised the tower. Five water towers were built on Gorter's modified-Hale design: four for San Francisco and one for Los Angeles. Out of 110 water towers used in North America, 44 were based on Hale's 1888 design.

In 1905, Frederic S. Seagrave adapted his spring aerial hoist to raising a 65ft water tower. He sold his first two through the Canadian firm that he and his son operated, W.E. Seagrave of Walkerville, Ontario. The first went to Winnipeg, Manitoba, in 1905, and the second to Montreal, Quebec, in

1906. Seagrave in Columbus, Ohio, built one for New York in 1907, and 10 more spring-raised water towers by 1930. Seagrave also built one hydraulically raised water tower for New York in 1929.

Like aerial ladder trucks, water towers carried more than just the heavy steel pipe that delivered a deluge of water 4 to 10 stories above street level. For example, in 1911, New York City's water towers carried the following impressive list of tools and equipment:

1 chipping hammer
1 monkey wrench
2 hubcap wrenches
1 hydrant wrench
2x4in hose spanners
2x3in hose spanners
2 cylinder rod stuffing box wrenches
2 spanners for Glazer nozzles
1 flat chisel
1 cylinder shifting bar
1 hand lamp
1 oil can
2 lengths of 4in hose
50ft of steam thawing hose
10 hose increasers, from 2.5 to 3in
4 hose reducers, from 3.5 to 3in
1x3in plug
Nozzle tips for mast: 1x2in, 1x1.75in, 1x1.5in
Nozzle tips for deck pipe: 1x2in, 1x1.75in, 1x1.5in

Even at their tallest, water towers could throw water only to a tenth-floor window. But many buildings exceeded that height. At the 1892 Columbian Exposition world's fair in Chicago, architects competed to build the biggest, tallest, and flashiest buildings at the fair. When the fair closed, architects continued to outdo each other. By 1913, New York faced the prospect of fires nearly 800 feet above Broadway in the new Woolworth Building. But in the 1890s, firefighters had a new tool to deal with that, too: the standpipe. A threaded hose connection was placed at the base of the skyscraper, and steamers could pump directly into it. A pipe ran up the full height of the building. At each floor, another pipe led off from this main standpipe, with a fire hose attached so it was easy and convenient to fight a fire on any floor. At various floors along the way to the top, built-in pumps boosted the water pressure to force water higher, up to the top floor if need be.

Fire Chief George C. Hale of Kansas City, MO, whose chemical and hydraulic water towers proved so popular, contributed another innovation to the fire service: the automatic horse harness. The harness was suspended from the firehouse ceiling. When the fire alarm sounded in the fire station, it automatically opened the doors to the horse stables. The trained fire horses then took their assigned places in front of the steamer, hose wagon, ladder truck, or water tower. The harness rigging automatically dropped over each horse. A firefighter snapped the harness onto each horse, and the apparatus was out the door in less than a minute after the fire station received the alarm.

One more new type of fire engine gained a foothold in the 1880s and 1890s: the fire boat. The first fire boat appeared in New York in 1800, when volunteer firefighters mounted a hand pumper on the deck of a wooden barge to create what they dubbed the Floating Engine Company. It could draw a few hundred gallons per minute from a river or the ocean, and supply it to hand pumpers on land who were fighting fires in piers or waterfront warehouses.

At the end of the Civil War, in 1865, New York switched from a volunteer to a paid fire department. In 1866, F.D.N.Y leased a steam tugboat, the John Fuller, and mounted two 1,000 gallon per minute pumps from Amoskeag steamers.

In the late 19th and early 20th century, two-wheel, hose reels, called "hose-jumpers," found favor with small-town fire departments, in outlying areas of big cities, and in industrial fire brigades. This fine example reposes in the New York Fire Museum.

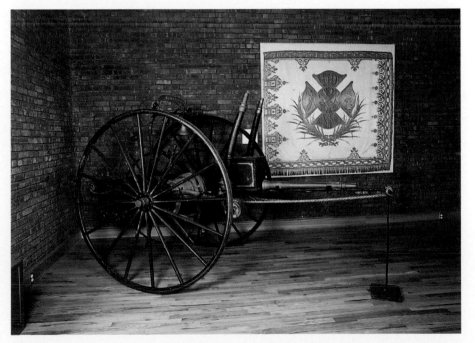

St. Paul, MN, bought this Seagrave spring raised water tower in 1917. It used the same spring lifting cylinders as Seagrave's aerial ladder trucks. *Seagrave Corporation*

Despite these firsts, Boston, not New York, would place in service the first boat specifically built for firefighting. The William M. Flanders, placed in service in 1872, was 75ft long, and could pump 2,500 gallons per minute.

New York's placed its first real fireboat, the William F. Havemeyer, in service in 1875. 106ft long, it could pump 2,870 gallons of fresh or salt water per minute.

By 1900, no fewer than 36 fireboats, with capacities exceeding 1,000 gallons per minute, were serving America's fire departments. In 1911, a typical New York City fireboat carried the following equipment:

1x1.75in and 1x2in open tip nozzle, for 3.5in hose
1.75in, 2in, 2.25in, and 2.5in nozzle tips to fit 3in deck pipes
1x3.5in to 2.5in and 1x3.5in to 3in hose reducer
2.5in, 3in, and 3.5in rail pipes
1x3in to 3.5in hose increaser
2x3.5in pipe holders
Siamese connection to join 2x3in hoses into 1x3.5in hose
1 gated three-way connection to split a 3.5in hose into 3x2.5in lines
1x3.5in to 3in and 1x3in double female connection
1 hose jacket for 3.5in hose
1 high-pressure Paradox pipe holder
1 two-way gate connection, 4.5in to 3in
1 single connection with gauge for 3in outlet; one swivel connection reducer, 3.5in to 2.5in
1 combination pressure regulator with gauges
1 hole drift
1.75in and 1.5in open nozzles for 3in hose
1.375in and 1.5in nozzles for 2.5in hose

Compare the equipment of this water-based engine companies, to the equipment carried by a typical land-based engine company of this era, consisting of a horse-drawn steamer and a horse-drawn hose wagon. Again, this list is as used in New York City in 1911:

2x4.5in hard suction hoses, 10ft 6in long
1x4.5in soft suction hose, 4ft long
1x4.5in suction swivel
1x4.5in x 2.5in reducing suction swivel
hydrant connection
portable hydrant nipple
fresh water hose and connection
steam thawing hose
hydrant pump
suction basket and rope
2 play pipes
squirt oil can
chipping hammer and flat chisel
monkey, combination, Stillson, alligator, socket, two adjustable, and three open-end wrenches
2 single-end and 1 double-end hydrant wrenches
suction wrench
high-pressure hydrant wrench
hub cap wrench
union nut wrench
four stuffing-box wrenches
shovel, poker, slice bar, and starting bar
driver's seat cushion and safety strap
tool kit for pump packing.

Despite all these new types of tools and equipment, however, the era of the Great Fire was not yet over. In 1904, it was Baltimore's turn.

The Great Baltimore fire began on Sunday morning, February 7, 1904, in the basement of a dry goods store. The first due Fire Company, Engine 15, arrived just 49 seconds after the first alarm was sent. Form what they could see, the firefighters considered it such an insignificant fire, that they used the one-inch rubber hose from their soda-and-acid chemical tank on their hose wagon, not even bothering to hook up hoses to the steamer.

But the burning rubber on which their stream sprayed included discarded celluloid novelties that burst into flames, shooting up an elevator shaft. Seven minutes later, these flames set off an explosion on the top floor, blowing out all windows in that building, and in many other nearby buildings.

As in the Great Chicago fire of 1871, a strong wind was blowing, and it spread the flames across Baltimore's business district. As would occur in San Francisco two years later, this fire would early on claim the chief of the fire department. Shortly after his arrival at this conflagration, Baltimore Fire Department's chief engineer, George W. Horton, was struck by a falling power cable and electrocuted. Without their leader, Baltimore firefighters were not as effective at controlling the fire in its early stages as they might otherwise have been. By the time the firefighting effort was seriously underway, the rapidly-spreading blaze had reached temperatures as high as 280°F.

Late Monday afternoon, 30 hours after the fire started, 36 fire companies made a last-ditch stand at a polluted, smelly, little creek. But their efforts halted the fire's advance, and it died out for lack of fuel. When it was over, 140 acres, about 80 city blocks, lay in ruins, with $100 million in property damage, and 40 firefighters injured although miraculously none were killed.

As in previous Great Fires, such as Chicago in 1871, the Baltimore blaze taught some valuable lessons. One was that with new man-made materials, such as celluloid film, firefighters had to be prepared for new fire dangers they had never faced before.

A surprising lesson occurred because various cities sent firefighters and fire engines to assist Baltimore, even from as far away as New York. Every American city had developed its own hose thread. That is, the pitch (angle) of the screw threads on hose couplings, and even the number of threads per inch, varied from one city to the next. This meant that the fire hoses that fit New York's steamers did not fit Baltimore's fire hydrants,

and dozens of fire engines sat idle for inability to connect to a water source. A few out-of-town engine companies drafted from Baltimore harbor or set up barricades around open hydrants to create makeshift reservoirs. But when 152 barrels of whiskey caught fire, and sent the burning alcohol into the streets, not only did these now-alcohol-laden reservoirs become useless against fire, but also they ignited and destroyed three steamers that relied upon them.

As a result of the Great Baltimore Fire of 1904, the National Board of Fire Underwriters developed a National Standard hose thread, and most cities have long since adopted this standard. In the huge wildland fires that annually befall the Los Angeles area, fire departments from all over California can and do send men and fire engines to help, confident that their hoses with National Standard couplings will couple to fire hydrants, and to the hoses of other fire departments. Amazingly, over 90 years later, there are still some fire departments in America that stubbornly cling to their own hose threads, and still have not adopted the National Standard thread.

As America's fire service entered the 20th century, its arsenal of fire engines and firefighting equipment had grown far more complex and varied than the simple bucket, hook, ladder, swab mop, and wood rattle that were the 17th century firefighter's only tools. And another major change was about to occur: the motor fire apparatus.

Many fire stations still in use in New York City today have reached the century mark. You can still see the horse stalls and haylofts inside. New York real estate has become simply too valuable to tear down the old stations and put up new ones. Some careful observers can thus find an unexpected architectural touch from the days before concrete and glass became the norm, such as this firefighter gargoyle on the quarters of Engine 24 and Truck 5 in Greenwich Village.

CHAPTER FIVE

Fire Equipment in the Early Motor Era 1910-1940

In the early 1900s, just as the motor fire apparatus era was dawning, America's fire engine industry underwent major changes. In 1891, the four largest steamer manufacturers, Ahrens, Silsby, Button, and Clapp & Jones, had merged to form American Fire Engine Co. In 1900, most of the other major fire engine manufacturers merged to form International Fire Engine Co. This new company included LaFrance (makers of steamers and ladder trucks), manning (chemical engines and hose wagons), Fire Extinguisher Manufacturing Co. (itself the merger of the Babcock, Champion, and Hale fire apparatus companies), Holloway (chemical engines), Gleason & Bailey (hand pumpers and ladder trucks), and Macomber (chemical engines and hose wagons).

In 1904, American Fire Engine Co. and International Fire Engine Co., already owned by the same group of investors, merged to form American-LaFrance Fire Engine Co. The management of Ahrens Manufacturing Co. did not like this "fire engine trust" and withdrew their company from the new conglomerate. When American-LaFrance threatened to sue them for using the Ahrens name, they briefly named their company Cincinnati Engine & Pump Works, but in 1905, won the right in court to call their firm Ahrens Fire Engine Co. In 1910, Ahrens was renamed Ahrens-Fox. The stage was now set for American-LaFrance and Ahrens-Fox to become fierce rivals in the new motor fire apparatus industry.

The first use of gasoline power in the fire service was in 1886 in St. Paul, MN, only a year after Gottlieb Daimler and Carl Benz independently introduced the world's first gasoline-powered automobiles in Germany. But unlike the automobile, the fire service's first use of gasoline was not for vehicle propulsion. A new 900 gallon per minute Ahrens steamer for Engine Company 3 had a gasoline-powered heater attached to the outside of its boiler, to keep the boiler tubes inside hot at all times. Steam could thus be generated more rapidly than if the boiler started from cold water.

Equipment carried on a steamer had always been minimal. Two 10ft long hard suction hoses were usually mounted along both sides of the engine. These consisted of steel springs encased in rubber, with brass couplings on the end, male thread on one end and female on the other. The suction couplings were generally nickel-plated. Two Dietz Fire King lanterns, either brass or nickel, hung from the sides of the boiler. At the rear of the boiler, in front of the engineer's platform, were the simple tools of the engineer's trade: an oil can for lubricating the steamer's moving parts, a shovel to feed coal into the fire box below the boiler, and a poker to stir-up the fire so it stayed hot. Suspended from chains under the firebox was a grate pan, to catch the hot ashes from the boiler before they could fall in the grass and start a fire — this pan was removed at the fire station, to dispose of these ashes.

In 1892, Charles H. Fox of American Fire Engine Company made the steamer slightly more complicated by adding a wooden hose box to the front end of the steamer, between the pump and the horses, and under the driver. Because the first of these were used at the World's Columbian Exposition in Chicago, this combination steamer and hose wagon was promptly dubbed the "Columbian engine." They never gained favor, however, perhaps because they did not really replace hose wagons. Other than the hose itself, Columbian engines carried none of the equipment commonly found on hose wagons of that day, such as fire extinguishers, extra nozzles, rope, axes, small ladders, and in some cases, even soda-and-acid chemical tanks.

With the coming of the motorized era, however, the functions of an engine and a hose wagon really could be combined on one chassis. As a result, the list of equipment carried on an engine company grew much longer, as we will see.

The first gasoline-propelled fire engines, however, were not pumpers.

In 1902, the Underwriters Salvage Corps of Cincinnati, a private firefighting squad run by local insurance interests, obtained a Winton automobile with two-cylinder gasoline engine, and had a local bicycle shop install seats at the rear for extra manpower, and wicker picnic-type baskets on the sides. This is believed to be the first use of a gasoline-propelled fire engine in the American fire service, and possibly in the world.

The baskets along both sides of its rear body carried heavy canvas salvage covers, which the crew draped over valuable merchandise in factories and warehouses, to protect these goods from fire and water damage. The baskets also carried hand tools, such as rakes, shovels, and hoes, used in post-fire overhaul of buildings and their contents. These wicker baskets had a nasty habit of falling off while turning sharp corners, and in 1909, Cincinnati's own Ahrens Fire Engine Company replaced them with bolted-on steel equipment lockers.

Seagrave of Columbus, Ohio, was the first fire engine manufacturer to offer gasoline-propelled fire engines as a regular part of their product line.

In 1902, Lee A. Frayer and William

"THE FLYING CHIEFTAIN"

An Automobile for Chiefs of Fire and Police Departments

THE SEAGRAVE COMPANY   COLUMBUS, OHIO.

Miller, began developing automobile engines in Columbus, for use in cars and trucks. Frayer-Miller made their first complete automobile in 1904, and America's first six-cylinder engine, an air-cooled 36-hp model, in 1905. In 1906, Frayer-Miller racing cars won every race they entered, and three of them completed the grueling Vanderbilt Cup race, where merely finishing without a breakdown was a triumph.

In 1907, Frayer-Miller approached Seagrave, which was already selling Frayer-Miller automobiles for use as chief's cars, about using their commercial truck line for Seagrave fire engines. Impressed with the performance of the Frayer-Miller engine, Seagrave promptly built a chemical and hose wagon on the flat front, or "buckboard" Frayer-Miller truck chassis. An air-cooled Frayer-Miller engine, mounted behind the driver powered it.

Above: This Seagrave ad, from the September 9, 1905, issue of *Fire & Water Engineering* magazine, offered Frayer-Miller automobiles adapted for use as fire chief's cars. *Ed Hass collection*

Below: Sonoma, CA, still owns this 1917 Nash chemical and hose cart. Bodywork is probably by Peter Pirsch & Sons, based in Kenosha, WI. *Ed Hass*

On June 27, 1907, Charles F. Kieser, superintendent of the Seagrave factory, took four local fire chiefs on a 105-mile drive aboard this rig, from Columbus to Chillicothe, Ohio, and back. The entire trip took 8h 17min, averaging 17mph.

In 1908, Chicago placed the prototype motorized Seagrave in service for a 30-day trial period, but decided that the chassis was not strong enough for fire duty. Chicago Fire Department did not buy any motor fire apparatus for another three years.

Before Chicago tested this prototype, however, Seagrave made several other road trips with it, all in the range of 60 to 90 miles. As a result of those tests, they made some design changes for their second automobile fire engine. They switched to a larger clutch, increased the air flow for cooling, and lengthened the wheelbase to carry more equipment.

Seagrave sold their second motorized fire engine, with these improvements, to Vancouver, BC, Canada, in 1907. It was also a chemical and hose car. Vancouver was so pleased with their Seagrave that, in 1908, they bought two more chemical and hose cars, and a motorized Seagrave aerial ladder truck. Unlike the chemical cars, the tractor of this aerial ladder truck placed the air-cooled Frayer-Miller engine in front of the driver.

In 1909, Frayer-Miller filed bankruptcy, and Seagrave purchased the manufacturing rights to their larger engines, as well as most of the company's tooling. They also acquired the remaining stock of unsold automobiles, which Seagrave fitted with searchlights, Dietz Fire King lanterns, and fire extinguishers, and resold as fire chief's cars.

The Seagrave buckboard, with its midship-mount, air-cooled engine, quickly gained popularity, and remained in production through 1913. At least three Seagrave buckboards survive today, a 1909, a 1911, and a 1913. But by 1911, Seagrave had also adapted the front-end sheet metal (radiator, hood, and fenders) of the front-engine Frayer-Miller automobile, to fire engines.

Also in 1911, Seagrave offered its first water-cooled engine. Air-cooled engines simply could not dissipate enough heat for the demands of a fire engine, which often sat idling at a fire for hours, with no air circulation to cool the engine. Water-cooled engines use a combination of air flow and circulating cold water to cool the engine.

Also in 1911, Seagrave teamed up with Gorham Fire Apparatus Company of Oakland, CA, to offer the first fire engine with a centrifugal pump. Seagrave designed and built a special chassis and unique six-sided hood and radiator for it, and Gorham

The first Ahrens-Fox motor fire engine undergoes a road test at Cincinnati in the autumn of 1911. It was the first fire engine to use the same gasoline motor to power both the driving wheels and the pump, and to carry pump and hose on the same chassis. After fitting it with a larger pump and other mechanical improvements, Ahrens-Fox sold this pumper to Rockford, IL, in December, 1911. *Ahrens-Fox Company photo via Steve Hagy*

Lenox, MA, still owns the first motorized American-LaFrance fire engine. The 1910 chemical and hose car is ALF Serial #1. *Ed Hass*

installed the pump. Tested in Oakland on June 27, 1912, it pumped 850 gallons per minute for a steady three hours. Several of these Seagrave-Gorham pumpers were built up to 1915. All were strictly pumpers, and required a second piece of apparatus to carry the hose. At least one Seagrave-Gorham pumper survives today, a 1913 at the automobile museum of the Imperial Palace Hotel in Las Vegas, NV. In 1914, Seagrave began offering pumpers with the company's own centrifugal pump, and these did carry their own hose.

Seagrave was not the only fire engine manufacturer to switch to automobile fire engines in those early years; they were just the first. On August 22, 1906, Ahrens-Fox of Cincinnati patented a motor fire engine that mounted a six-cylinder piston pump above a six-cylinder gasoline motor, with a connecting rod linking each pump piston to each engine cylinder. It was basically a multi-cylinder version of the steam fire engine pump, where the steam cylinder was joined to the pump piston. Having the pump in constant motion, even when the engine was racing to a fire, would have wasted a lot of power and fuel, and Ahrens-Fox evidently never built an engine to this patent.

In 1910, Ahrens-Fox began working on its first motor fire engine. Unlike Seagrave, which learned its lessons with motorized hose wagons first before adding the complication of an engine power take-off for a pump, Ahrens-Fox jumped right in with both feet. As Seagrave had four years earlier, Ahrens-Fox found that its prototype left much to be desired, and after demonstrating it in the autumn of 1911, they rebuilt it with a larger piston pump and several other improvements.

As with Seagrave, the first motorized Ahrens-Foxes used someone else's power plant, in this case from Hirschell-Spillman of

Ramon Kehrhahn of Connecticut, one of the few craftsmen today skilled in the gold leaf art, restored this Ahrens-Fox piston pumper, Model M-K-3 #812, originally used in Piqua, OH. The only non-authentic feature is the use of modern tires. *Ed Hass*

Tonawanda, NY, but by 1914, Ahrens-Fox Company offered front-mount piston pumpers with a six-cylinder engine of their own design.

American-LaFrance, too, joined the motor fire apparatus craze, building two hose wagons on Simplex automobile chassis as early as 1909, and producing the first motorized hose wagon on their own chassis for Lenox, MA, in 1910. The first motorized American-LaFrance pumper and hose wagon combination, with rotary gear pump, followed in 1912.

By the 1920s, American-LaFrance (ALF), Seagrave, and Ahrens-Fox were the three biggest manufacturers of motorized fire engines, with Seagrave's sales volume about two-thirds, and Ahrens-Fox's about one-half, that of ALF. A fierce competition grew between these three manufacturers, over which was the best type of fire pump.

Ahrens-Fox's piston pump was the fastest to get water, and at a fire every second counted. The Ahrens-Fox pump could also pump for long periods under the most extreme firefighting conditions, such as long hose lays and high elevations. But the piston pump had the greatest number of moving parts that could wear out, and it was expensive to manufacture and to maintain.

The American-LaFrance rotary gear pump was nearly as fast as the piston pump, but it had fewer moving parts to wear out, and its lighter weight meant it could be mounted on smaller chassis for rural as well as big-city firefighting.

The Seagrave centrifugal pump took the longest time of the three to throw water. Unlike the piston and rotary gear pumps, a centrifugal pump was not positive displacement. That meant it had to have a separate, small rotary-gear priming pump, to start the main centrifugal pump so it could draw water from a river, lake, or hydrant.

But the centrifugal pump had the fewest moving parts, so it was easier and cheaper to manufacture and maintain. Its main chamber used centrifugal force to throw large volumes of water at great pressure, and making the chamber bigger increased the gallons per minute capacity of he pump. There was a practical upper limit to the diameter of a pump piston, or the size of a gear in a rotary pump, before the weight involved outstripped the practical limits of the power required to run it. But the centrifugal pump seemed to have no upper limit for capacity, as evidenced by the 20,000 gallon per minute centrifugal fire pump that Rotterdam, Holland, placed in service in the early 1990s. Since the 1950s, every new fire engine built and sold in America has used a centrifugal pump.

Although American-LaFrance, Seagrave, and Ahrens-Fox would dominate the motor fire engine industry in the 1920s, none of them were the first to introduce a motorized pumper. That honor went to Waterous Engine Works of St. Paul, MN, which delivered America's first motorized pumper to Radnor Fire Company of Wayne, PA, in 1906. Waterous had been making horse-drawn pumpers, whose pumps were powered by gasoline engines, since 1898. The engine for Wayne had a second gasoline motor to propel it through the streets to the fire, and like the Seagrave-Gorham four years later, it carried no hose. It was not until Ahrens-Fox developed its first motor fire engine in 1911, that a single gasoline motor propelled a combination pumper and hose wagon to a fire, and then could be switched to using the same engine to power the pump.

Mack Trucks of Allentown, PA, made its first two motorized fire engines in 1911: a service ladder truck for Morristown, NJ, followed soon after by a rotary gear pumper and hose car for Union Fire Association of Lower Merion, PA. Mack would be the fourth-largest fire engine manufacturer by the 1920s, and by the 1960s, was second in

The days of the flat-brimmed firefighter's uniform cap were numbered by the time B. Pasquale & Co. of San Francisco, CA, made this cap in 1910. By 1915, the curved-brim cap was replacing this style all over America. Other fire apparatus enthusiasts might recognize this cap, as the author wears it to fire engine meets all over the U.S.

sales volume only to American-LaFrance.

With the motor fire engine, firefighters finally had a way to respond to a fire quickly, and to deliver large volumes of water at high pressures. After the earthquake-induced fire that burned down San Francisco in 1906, there would be no more Great Fires that burned down entire cities.

This is not to say that there would be no more fires causing heavy damage and great loss of life. But from now on, these firs would be confined to one building, or to one city block, not sweeping 100 or more city blocks.

One such major disaster befell New York City in 1911. At that time, most of the city's steamers, ladder trucks, and water towers

Around 1910, a central fire alarm telegraph office received fire alarms from all over the city, and then used this massive console to reroute it to the appropriate fire stations. This display is at New York City Fire Museum.

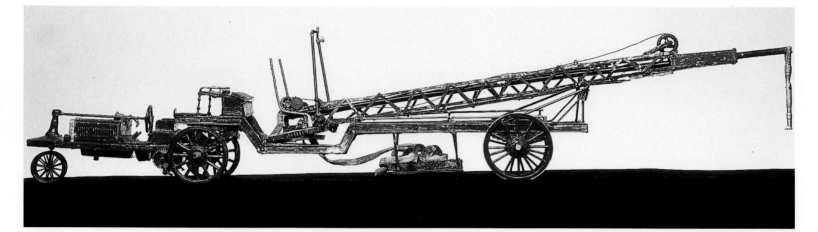

Bill Hass, the author's father, is a leading expert on water tower fire apparatus. In the 1950s and 1960s, he hand-built scale models of several horse-drawn and motorized water towers, and I am fortunate to have a few of these in my collection. My favorite, however, is this 1894 Champion water tower with 1911 Knox-Martin three-wheel tractor. The actual machine was used in Springfield, MA. Knox-Martin was a marriage of Knox Automobile Co. of Springfield, MA, and Martin Carriage Works of York, PA. Sadly, the real fire engine, on which this model is based, was junked in 1955. Today, only one actual Knox-Martin "tricycle" tractor survives, at San Jose, CA, only 10 miles from my home in Sunnyvale, CA. So I get the double treat of having a model of one of these odd tractors in my living room, and the real thing a short drive away.

were still horse drawn. On a very few rigs, F.D.N.Y. had replaced the front wheels and horse hitch with a two-wheel or four-wheel motorized tractor, and a few fire engines built new with gasoline motors had recently been placed in service.

The Triangle Shirt Waist Company, a sweatshop garment factory occupying the top three floors of the ten-story Asch Building in New York's Washington Square. Most of the other companies in the building closed over the weekends, but Triangle's 600 employees, mostly young girls aged 14 to 20, were required to work a six-day week, Monday through Saturday, at subsistence wages.

The work tables, where these girls cut and sewed flimsy material into ladies blouses, were packed tightly together, with no aisles for easy exit in case of fire. Triangle's management kept the factory doors locked from the outside, so the girls would not try to sneak out for lunch or go home early, so again, the girls had no means of escape in case of fire. The Asch building was very old, and its single outside fire escape was rusty and in disrepair, so it, too, was not a reliable means of escape in case of fire. This combination of factors was a recipe for disaster.

That disaster would come at 4:45pm on Saturday, March 25, 1911, just as Triangle employees were preparing to go home. Someone spotted a fire smoldering in a rag bin on the eighth floor. By the time a bucket could be filled with water and tossed on the rags, the flames had spread to the tissue paper and cloth scraps scattered on the wooden work tables and wooden floors.

One of the two exit doors on the 8th floor led to a stairway. But management had locked this door to prevent the girls from taking breaks.

Mercifully, Triangle workers found one unlocked door through which they could escape. But the hallway had been built only wide enough for one person at a time, so that management could check the girls' purses to make sure they were not stealing blouses and taking them home to wear. Only a few had time to escape by this route. Others were able to exit by the one fire escape, before it gave out from the heat and its general state of dilapidation. Despite these impediments, most of the 8th floor workers escaped unharmed.

Flames leaped out of the 8th floor windows, spreading upward to the 9th floor. There, the fire ignited the completed blouses hanging from racks. The 9th floor workers had even less time to escape than those on the 8th floor. One of the strong male employees forced open the locked stairway door, but the stairway was already too full of smoke for all but the bravest to venture out that way. The 9th floor Triangle employees also faced the narrow, one person at a time hallway, and few escaped that way.

Some managed to cram into an elevator and ride down to the ground floor. Before the elevator could return, the heat and smoke became unbearable, and some 30 desperate girls flung themselves down the elevator shaft, falsely hoping the elevator would catch them unharmed. The weight of piled-up bodies prevented the elevator from returning to rescue those still trapped.

A very few managed to escape down the building's only fire escape, before the heat of the fire and the weight of those fleeing the 8th and 9th floors collapsed the escape.

Then 60 more girls climbed onto the wind ledges, but when the flames behind them began to singe their backs, they leaped to their deaths rather than slowly roast.

Firefighters liked fancy decorations on their fire engines, and on their formal uniforms. These delegates' ribbons, worn military-style on the left side of the uniform jacket, are from conventions of the New York State Firemen's Association. At left, from the 1913 convention at Newburgh, NY. At right, from the 1914 convention in Binghamton, NY. Both ribbons are displayed at the New York City Fire Museum.

FIRE EQUIPMENT

A study in contrasts is this pair of fire hydrants at the San Jose Historical Museum, in Kelley Park, San Jose, CA. The large red double hydrant is displayed as an antique in front of the Trolley Barn, where antique trolley cars are restored and maintained. The newer and smaller yellow single hydrant is the one that San Jose Fire Department would use today if that same Trolley Barn were to catch fire.

Above: Indoor fire alarm pull station #3-937-4, and outdoor alarm box 113, both saw use in New York City in the 1920s. They are now displayed at the New York City Fire Museum.

Right: State of the art in fire hose, as advertised in the June 1, 1927, *Fire Engineering* magazine, featured inner rubber tube, surrounded by cabled inner jacket, and woven cotton outer jacket. Outer jacket was coated with wax (as in candles) and para gum (as in turpentine) to protect against mildew and rot. These flammable coatings were probably not the best chemicals

Firefighters tried to catch them in the life nets, but the 100ft leap made the dead weight of each girl equivalent to two tons.

Firefighters could do little to battle the flames on the 8th and 9th floors, or to rescue the workers there. The aerial ladders reached only to the 6th floor, and the water

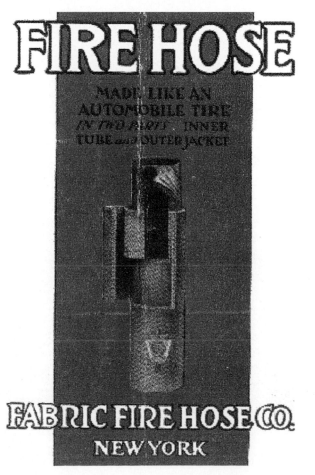

towers only to the 7th. Firefighters dashed up the stairs with their hoses, and within 18 minutes after their arrival, the fire was out. But despite the efficiency of these firefighters speed, 146 workers died before the fire could be extinguished.

As in the various Great Fires of the previous 200 years, lessons learned in this fire were applied to prevent a repeat of the disaster.

New York enacted 30 new fire ordinances as a result of this fire, and most other cities followed their lead. Automatic sprinklers were required on all floors of all factories and office buildings. Multi-story commercial buildings were required to have wide hallways and at least three internal stairways. The stairway doors were to remain unlocked at all times. These buildings were required to have at least two exterior fire escapes on each side.

Commercial buildings were to be inspected by city firefighters on a regular basis, to make sure that means of escaping from fire were accessible and in good repair. Combustibles, such as rags, were to be stored in designated, fire-resistant cabinets or lockers, not strewn about as at the Triangle factory. Rooms were to have maximum occupancy figures, assigned and posted by fire inspectors. There would be no more crowding of work tables and people into tiny spaces, no more rooms occupied by too many people to escape using available exit routes, in the short time before flames and smoke would overtake them.

One other result of this fire was the formation of the Ladies Garment Workers union. Using work stoppages and strikes, and supported by the public outcry over the deaths of 146 young women, the union quickly gained power throughout the garment industry. This union abolished the sweat shop conditions that kept workers virtual prisoners six days a week, distrusted and searched by their employers, and paid salaries that could barely support their food and rent. In short, garment factory workers throughout the U.S. were accorded their basic human rights after this fire.

Fire Alarm Cards

Fire alarm cards were used to help central fire alarm dispatch officers send the best-placed stations to the scene of the fire. The

The Big 4 fire engine manufacturers of the 1920s (American-LaFrance, Seagrave, Ahrens-Fox, and Mack) all equipped their fire engines with Dietz Fire King hand lanterns, made in Syracuse, NY. This one, in the author's collection, has been cleaned-up, but still needs new nickel plating. Typically, a restored Dietz lantern costs $300 to $500 in today's collectors market, and nearly everyone restoring a 1920s vintage fire engine seeks a pair of them.

Built in the 1920s, New York City fire alarm box #1376 is now on display at New York City Fire Museum.

example, take the fire alarm card for Fire Alarm Box #1395, at 83rd Street and Cottage Grove Avenue in Chicago. It would have provided the following step-by-step instructions. On the first alarm, the dispatcher retransmitted the alarm to Engine Companies 47, 72, and 82, Truck 42, Squad 5, and the Chief of Battalion 19.

If Battalion Chief 19 determined that the fire was serious enough to merit a second alarm, the dispatcher would summon Engines 63, 73, and 122, Truck 34, Squad 12, High Pressure hose wagon 6, Ambulance 5, District Chief 5, and Battalion Chief 14. He also would have moved Engine 9 to cover for any calls in Engine 63's neighborhood, Engine 29 to Engine 47, Engine 75 to 73, Engine 80 to 82, Truck 37 to Truck 34, Trucks 40 to 42, Battalion Chief 16 to Battalion 14, and Battalion Chief 9 to 19.

From this second alarm on, District Chief 5 would be in command of the fire scene. If he decided that the fire required a third alarm, the dispatcher sent Engines 51, 54, and 100, Truck 20, Squad 13, and Water Tower 3 to the fire. He also moved Engine 6 to cover for Engine 51, Engine 28 to 54, Engine 48 to 72, and 52 to 122.

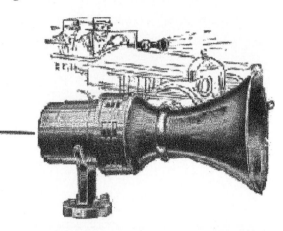

Left: In the early automotive era, many states, such as New Jersey, outlawed sirens on emergency vehicles, such as this one made in 1927 by Sterling Siren of Rochester, NY. The reason: they scared the horses! *Ed Hass Collection*

Below: The Hall of Flame, Phoenix, AZ, owns this 1915 Brockway chemical and hose car, ex-Kutztown, PA.

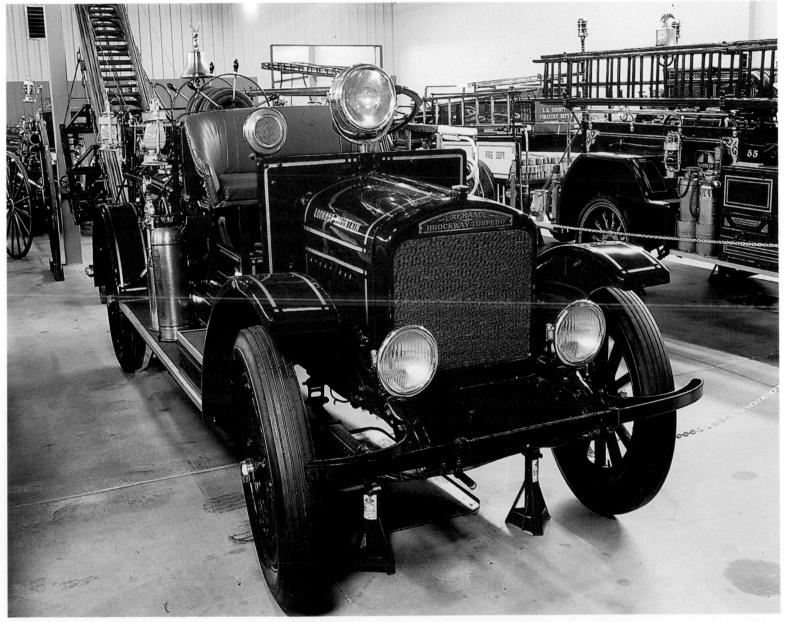

Right: The *Detroit News* photographed the crew of Engine 9, on Riopelle and East Larned Streets in Detroit, MI, aboard their brand-new Ahrens-Fox #981 in October 1921. *Courtesy Detroit Fire Dept, Historian Clarence Woodard*

Below: Arthur Bennett of Fair Haven, NJ, owns this restored 1922 Ahrens-Fox Model K-S-4 #1126. Paint color, gold leaf fire equipment, and even button-tufted black leather seats, match the original appearance at Hampton Beach, NH. Only the tires are modern, the original solids long replaced. *Arthur Bennett*

Above: Beverly Hills Fire Department American-LaFrance vehicle.

Left: In February 1928, demolition workers dismantling the aging Pocassett Mill in Fall River, MA, first removed the overhead fire sprinkler pipes, then cut up the rest of the structure, using acetylene torches. The predictable result was a blaze that took days to extinguish. Here, firefighters assisting Fall River, hastily remove ice from their 1917 Ahrens-Fox piston pumper, so that they can move it. *Detroit Fire Dept Historian Clarence Woodard Collection*

Tough looking and durable, this circa 1935 USA pumper of Arcadia, CA, still operates frequently at antique fire engine events throughout California. *Ed Hass*

A fourth alarm would send five more pumpers to the scene: Engines 60, 61, 101, 126, and 129. It would also have sent Battalion Chief 29 to the scene. Relocations would have been Engine 21 to 126, 98 to 60, and 102 to 29. Battalion Chief 8 would cover for Battalion 12.

A fifth alarm sent Engines 9, 19, 45, 52, and 87 to the fire. It also moved Engine 12 to 48, 17 to 63, and 42 to 45.

Todat this would be an antique fire engine enthusiast's dream! There would now be 19 pumpers (all either Ahrens-Fox piston pumpers or chain-drive Seagrave centrifual pumpers), 3 ladder trucks (classic Mack Bulldog and Seagrave Suburbabite-style), 3 squads (probably Mack Bulldogs), two high pressure hose wagons, achain-drive Seagrave water tower, an ambulance, three cattalion chief cars, and a district chief's car — in total 33 fire engines.

Beyond a fifth alarm, Chicago dispatchers might special call specific apparatus, such as a fireboat, or call a general alarm that summoned every available fire engine and every city firefighter, on or off duty. At the 1934 Stockyards fire, and again at the 1939

Rosenbaum Grain Elevator fire (both were general alarms), Chicago sent over 100 pumpers, more than a dozen ladder trucks, all 3 Seagrave water towers, fireboats, ambulances, high pressure hose wagons, battalion and district chiefs, and so on.

The Emergency Squad

A new breed of fire engine appeared in America in the 1930s: the Emergency Squad, or heavy-rescue truck. True, big cities such as Detroit ran motorized emergency squads as early as 1908, but these were usually little more than a means of carrying extra manpower to a fire. If they contained any medical equipment at all, it was usually little more than a rudimentary first-aid kit and a stretcher without wheels, on which two men could hand-carry a fire victim.

The new breed of rescue truck, introduced in the 1930s, was designed to deal with emergencies of which earlier firefighters never dreamed: automobile accidents, airplane crashes, train derailments, gas-main ruptures, building collapses, and overly-curious children trapped inside refrigerators. American firefighters were now expected to offer medical aid not only to fire victims, but also victims of tornadoes, hurricanes, floods, earthquakes, and volcanoes. The new breed of rescue truck required its own electric generator, not only to light-up the scene at night fires, but to power a whole host of newly-invented electrical tools that helped rescue trapped victims from all of these types of disasters.

A typical rescue truck of the 1930s was an Ahrens-Fox Model Y-8 combination Floodlighting, Rescue, and Salvage Car, shipped to Mount Vernon, N.Y., on April 10, 1937. The 10ft long rescue body, mounted on Ahrens-Fox's 162in wheelbase chassis, was placed in service at the quarters of Engine 4, on Sixth Avenue, just 5 days later. Mount Vernon's Fire guided Commissioner William A. Hallahan and Chief John Gibson assisted Ahrens-Fox engineers in designing this vehicle. For the next 16 years, it responded to all box alarms in Mount Vernon, and was often special-called to various emergencies. A new rescue truck, made by Approved Fire Equipment of Rockville Centre, NY, took its place in 1953, but the Ahrens-Fox continued as a reserve rescue squad until junked in 1971. This 34-year service life is a testament to the high quality that went into every fire engine of the classic 1910-1940 era.

Just look at the long list of equipment carried on Mount Vernon's Ahrens-Fox rescue truck, as reported in the May, 1937, issue of Fire Engineering magazine:

"In a transverse compartment at the forward end of the body is mounted a fully-automatic Model K Kohler electric plant of 2,000-watt capacity, generating 110-volt D.C. Two 1,000-watt lights are permanently mounted on the top of the apparatus, and six 250-watt portable lights are carried in a center compartment, Two cable reels carry 1,000 feet of lighting cable on each reel.

"Thirty-six 12 x 18-foot salvage covers are carried in side compartments, and in addition to the floodlighting and salvage equipment, the following fire-fighting and rescue equipment is conveniently stored in compartments and mounted at accessible positions on running-boards and top of the apparatus body:

Left: New York's tall buildings and narrow alleys made placing conventional ground ladders difficult: they had to be placed at an angle and there often was not enough space between buildings to do that. The solution: the scaling ladder, which could hang vertically from a window or ledge. They had a single vertical beam in the center, with horizontal rungs protruding from either side. These are from the New York Fire Museum.

1 acetylene cutting outfit, complete
1 electric saw
1 life gun, in case
4 Burrell all-service gas masks
1 MSA hose mask, in case
2 McCaa gas masks
2 McCaa canisters
4 extra cylinders, McCaa masks
2 H&H inhalators
2 extra cylinders, inhalators
1 M.S.A. first aid kit
1 finger ring cutter
1 folding cot
4 asbestos blankets
1 wire and bolt cutter
2 red flags
11 cans Carboride
6 blankets
2 wading suits
4 electric heating pads
1 Voleske [iron] bar cutter
1 gas shut-off wrench
1 high-voltage tong
1 pair rubber gloves
2 Frigidaire keys
two 10-ton hydraulic jacks
2 extension lifts for hydraulic jacks

Below: The Burrell "all service gas mask" as seen in a contemporary advert. *Ed Hass Collection*

Right and Below right: American firefighters first started wearing identifying badges in the 1840s. By the time these ads for rival badge makers appeared on the same page of *Fire Engineering* magazine.

Below: Akron Brass of Wooster, Ohio, made these hose spanners, in the author's collection. The large, heavy brass wrench is from 1925, and had to be carried in the toolbox of a pumper. The small, lightweight aluminum wrench, made 65 years later, does the same job, but folds up to fit in a firefighter's pocket.

2 screw jacks
250ft of 0.5in rope
1x 24ft extension ladder
1x 12ft folding ladder
1x 10ft pike pole
1 crow bar
1 Detroit door opener
1 stretcher
1 New York type claw tool
1x12ft roof ladder with folding hooks
1x2.5 gallon S & A Extinguisher
2 CO2 extinguishers
1 CO2 transfer pump (charging unit)
2 Phomaire play pipes
1 2.5 gallon foam extinguisher
4x 1-quart C.T.C. extinguishers
4 electric lanterns
1x 2in auger
1x8in flat file
1 pick head axe
1 short handle flat shovel
1 short handle pitch fork
1 No. 10 sledge hammer
1x6in offset screw driver
1 alligator wrench.

Warning signals consist of locomotive bell, Federal-Coaster Electric siren, and Buckeye whistle. A 10in revolving searchlight is part of the apparatus' light equipment.

The list of tools and equipment carried on motorized pumpers was also considerably longer than on horse-drawn steamers. Every manufacturer had its own list of standard equipment, but all makes carried mostly the same equipment. Here, from a 1925 catalog

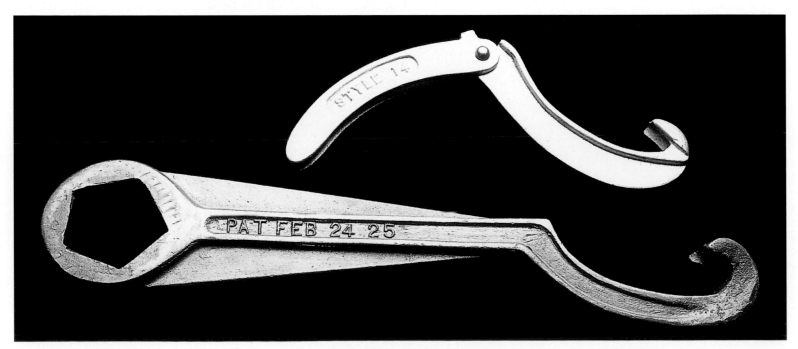

This Siamese connector allowed firefighters to split one large 2.5in hose line leading from a pumper, into two 1.5in lines leading into a building. This one, in the author's collection, bears the raised letters "S.F.F.D." and square hose threads that could only mean it was used in San Francisco, CA.

San Francisco used square, instead of traditional peaked, threads on hose couplings. This nozzle, made in 1928 and now in the author's collection, was later fitted with the adapter at right, to connect to conventional hose threads. Many older skyscrapers in San Francisco still have standpipes with square threads, fitted with similar adapters.

of Ahrens-Fox Fire Engine Company, is a list of the tools and equipment carried on a typical motorized pumper of that era:

Open end wrenches, socket wrenches, pliers, hammer, chisels, punches, screw drivers, rubber mallet. 5-gallon oil can. Grease gun. Axle, hubcap, magneto, and clutch wrenches. Pump stuffing box and valve stem wrenches. Suction and discharge hose spanner wrenches. Rubber hydrant-thawing hose. 6-ton lifting jack.

Steel hose body with wood slat floors, capacity for 1200 feet of 2.5in rubber lined hose with double cotton jackets. Two 6in hard suction hoses, each 10 feet long, mounted on steel troughs along both sides of hose body. Suction hose strainer, mounted on steel base on running board. One 20ft extension ladder with rope hoist and automatic lock; one 12ft roof ladder with steel folding hooks; both ladders made of solid-side, straight-grained fir. 8ft pole with drop-forged steel plaster hook.

Two 7lb, pick point fire axes, with blade sheaths and clasps. One 12lb crow bar of tool steel, with holders. Two Dietz Fire King fire department hand lanterns, made of brass, with nickel finish, mounted on hook brackets with steel clasps. Two Grether #2 battery-powered, electric hand lanterns. Two 3-gallon portable fire extinguishers, soda-and-acid type, nickel plated, mounted in sleeve retainers, on running boards. Two brass play pipes, with twine coverings, mounted on cone-shaped wood holders on rear step.

Leather splash guard, under body at rear. 60-gallon steel booster tank; 200ft of 1in rubber booster hose, on reel in rear step. 10in locomotive bell, at top center of dashboard. Lights and signals include two 10in Vesta drum-style headlights, one Vesta drum-style 10in searchlight, and one red tail light; Sterling hand-cranked siren, Buckeye exhaust whistle, and Mars swivel light optional.

Tractor-trailer aerial ladder trucks of the 1920s and 1930s and 1930s typically carried the following impressive list of tools and equipment, again from Ahrens-Fox literature of 1925 vintage:

Two 10in Vesta drum headlights; 10in Vesta drum searchlight; Vesta red tail light; trouble light; under hood light; speedometer light; Sireno electric horn. Tillerman's signal bell with floor-mounted control button; 10V electric signal bell for driver on dashboard. Four Dietz Fire King lanterns. Two wire equipment baskets on center platform.

Zerk 3-A grease gun. Axle nut, tire, hub cap, clutch, magneto, and aerial ladder air compressor wrenches. Open end wrenches, socket wrenches, pliers, hammer, chisels, punches, screw drivers. Gem hand oil can. Automatic truck jack with 48in folding handle. 5⁄16in, 3⁄8in, and 1⁄2in sockets. Five-gallon oil can for motor; 1-gallon Polarine oil can for aerial hoist.

75ft main aerial ladder, 32in wide, wood, in two sections, with pneumatic hoist and wood tackle block at top. Trussed wood ladders: 45ft extension x 24in wide, with tormentor poles; two 35ft wall x 24in wide; 30ft wall x 24in wide; 28ft wall x 20.25in wide; 24ft and 20ft wall x 19.5in wide. Solid-side wood ladders: 16ft wall x 16in wide; 12 foot roof x 15.75in wide, with steel folding hooks. 16ft wood crotch pole. Two 13ft plaster poles.

Four 7lb pick point fire axes; battering ram; four brooms; three crow bars; door opener. Two 3-tine hay forks; hose hoist; 6 cotton bale hooks; two mops (optional); hammer-head pick; 175ft of 3⁄4in manila rope; 2 rubber buckets (optional); 12lb, double-faced sledge hammer, with handle; two flat and two scoop shovels; 4 squeegees. Tin roof cutter with 30in spade handle; wire cutter with insulated handles. Two 3-gallon Childs soda-and-acid fire extinguishers.

One final change that the motor fire apparatus era brought, was standardizing the way in which fire engines and fire equipment were procured. Fire departments would advertise in publications such as Fire Engineering, or mail letters to fire apparatus manufacturers, inviting them to submit bids, and indicating the precise type of equipment they sought, whether a complete pumper or ladder truck, or a single firefighting tool such as fire hose. The various manufacturers who made products that met the fire department's specifications, would then send the fire department their standard printed literature, describing their equipment in great detail.

The manufacturers would also send a bid proposal, spelling out exactly what equip-

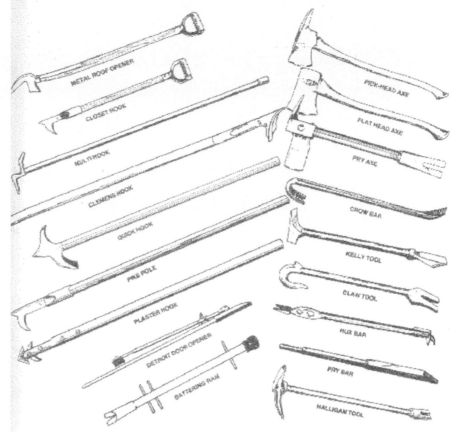

ment they intended to supply. The proposal showed how soon the town could expect delivery after the contract was signed (typically, a new pumper of the 1920s required 90 to120 working days to build). It also stated what the purchase price would be, and the terms of payment (for example, paid in full within 90 days after delivery). The bid proposal would also state whether a factory representative would train the firefighters in the use of this equipment, and if so, how many days the trainer would spend in that town.

The proposal even spelled out the guaranteed performance of the equipment. For example, a 1,000 gallon per minute pumper was expected to pump its full rated capacity at 120lb per square inch (psi) net pump pressure for six hours, half capacity at 200 psi for three hours, and one-third capacity at 250 psi for three hours. Failure to meet or exceed these guaranteed performance levels on delivery could be, and usually was, cause for the customer not to accept and purchase the new equipment. The manufacturer would then have to take the equipment back, either modifying it to meet the customer's requirements, or selling it to another fire department with less stringent requirements.

The new breed of heavy rescue truck, introduced in the 1930s, carried many of the tools seen here. In cities that did not have rescue trucks, and even in some that did, aerial ladder trucks would carry these tools. Some tools, such as axes, hooks, and pike poles, were very traditional. Others, such as the Detroit Door Opener and the Halligan Tool, had been only recently developed, to meet new firefighting needs unknown to previous generations.

1927 Ahrens-Fox Model J-S-41 service ladder truck # 1262 makes a rare appearance at a two-alarm fire in the H&H Dollar Store, 9215-21 Kercheval, Detroit, MI, on October 12, 1929, just two weeks before the Stock Market crash that led to the decade-long financial depression. *Courtesy Detroit Fire Dept, Historian Clarence Woodard*

Right: W.S. Darley of Chicago made or sold everything a firefighter needed, from complete fire engines to this 1930s curved-brim uniform cap. This cap, sporting crossed-axes insignia and lettered "California 231" belongs to Derek Lyon-McKeil of Sunnyvale, CA.

Far right: By the time this open-tip playpipe (in the author's collection) was made in the 1930s, shut-off nozzles were rapidly making this style of tip obsolete. The trouble with open-tip playpipes was that water could not be shut off quickly in an emergency, in case a hose line had to be moved quickly. Shutting off the pump rapidly was not practical as a way to close the hose line, because it created a tremendous "water hammer" that could burst a hose, shatter a nozzle, or send the firefighter on the nozzle end flying.

Armed with bids from several manufacturers, the mayor, city council, fire chief, and fire commissioners would, meet to award a contract to the bidder who most closely met the community's needs at the lowest purchase price.

Of course, sometimes the bids were stacked in favor of the town's favorite vendor. For example, Morristown, NJ, was partial to Ahrens-Fox pumpers, so in 1947, they called for bids to supply two pumpers with "front-mount, 6 cylinder piston pumps." Other than Ahrens-Fox, only one other fire engine manufacturer, Howe of Anderson, IN, still offered piston pumps in 1947. Howe's pumps had only two pistons, and were not front mounted.

As if the deck were not stacked heavily enough in favor of Ahrens-Fox getting this contract, Morristown required the successful bidder to accept a 1917 and a 1923 Ahrens-Fox piston pumper in trade as part payment on the new pumpers. Only Ahrens-Fox had a ready market of customers for purchase of second-hand Ahrens-Fox equipment.

In the Depression 1930s, many cities and towns were hard pressed to find money to replace aging fire engines, so they just kept patching the old ones up, and keeping them running far beyond their normally expected 20 year service life. In the few communities that were still buying new fire engines, competition between the fire engine salesmen was fierce. In those lean years, losing a contract to a competitor could literally mean no food on the salesman's table. One manufacturer had all sorts of shady schemes for winning contracts away from competitors. They would have local "citizens," who were actually their sales representatives, write editorials in the newspaper of the town considering a new fire engine, praising their own

Short, fat fire hydrant with rib-sided barrel and top has been a common sight on New York City streets since the 1930s. The last two decades of tight city budgets have let this hydrant severely rust, but it is still in use.

town was debating and awarding a contract, several of these huge salesmen would be present in force, to intimidate the solitary small, wiry salesman from each of their competitors. This was much the way the mere presence of a group of Mafia strongmen would intimidate those who testified against them, in racketeering trials of the 1950s and 1960s.

The 1930s were lean times for the politicians and fire chiefs, too. In many communities of the 1930s, fire apparatus salesmen could not close a contract without bribing the mayor, councilmen, fire chief, and board of fire commissioners. Salesmen could not afford to pay these bribes and kickbacks from their own pockets. And reputable manufacturers were not about to provide their salesmen with funds for such clearly illegal activities, no matter how desperately they needed sales in the roughest years of the Depression, 1931 to 1938. So ingenious ways were fond to hide these illegal but necessary costs of doing business.

One of the cleverest ways to hide these bribes and kickbacks was in providing one-off custom bodies, such as pumpers with custom enclosed cabs, or ladder trucks with fully-enclosed ladder racks. The manufacturer could then charge the customer, say, an extra $1,000 for the custom bodywork. Who was to say whether the extra sheet metal, and the extra labor to make a one-of-

Above: The Ahrens-Fox Fire Engine Company of Cincinnati, OH, cast the serial number of the truck in nearly every part of every fire engine they made. This "double female" adapter, in the author's collection, let firefighters connect the male ends of two hoses together, or the male end of a hose to male thread on a pump or fire hydrant. The serial number stamped in this one, 9004, reveals that it was from a 1935 pumper used in Hamilton, OH.

Right: In 1917, FDNY rebuilt this tractor, one of the 1913 fleet of American-LaFrance second size Metropolitan fire engines. The original engine was replaced by a more conventional Van Blerck motor.

brand and denouncing the competitor's equipment as "antiquated, obsolete, and not in accord with progress." Or they would befriend the local labor unions, pointing out that theirs was a union shop, and their competitor's was not. The local unions would then pressure the politicians to buy the union-made truck over the non-union truck, or risk losing the next election.

Another tactic of that manufacturer was hiring tall, muscular salesmen. When a

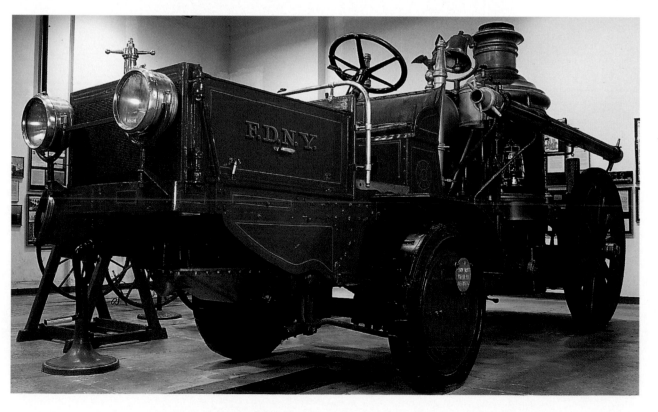

a-kind body, really cost $1,000. Or did it really cost only $500, and the city treasury paid the other $500 to reimburse the manufacturer for bribes to the various city officials who signed the contract for the new fire engine.

Sales of new fire engines picked up slightly in 1937, 1938, and 1939, but did not return to the levels they had enjoyed before the 1929 stock market crash. As the storm clouds of war drew over Europe and Asia in 1939-1940, America's entry into World War II would not be far off.

After Japan bombed Pearl Harbor on December 7, 1941, priority for the materials used in building fire engines (such as steel, glass, and rubber) went exclusively to the war effort. American-LaFrance, Seagrave, Mac, and Peter Pirsch made a very few fire engines during World War II, and those were mostly for the Army and Navy.

Most other fire engine manufacturers turned their plants over exclusively to production of military vehicles, weapons, and other machinery used by the military. For example, Ahrens-Fox built giant lathes for boring the holes in the big gun barrels on Navy ships.

Above: Unusual GMC cab-over-engine pumper, with pump mounted on front bumper. *Ed Hass*

Left: This Buckeye "Roto-Ray" light, originally used on one of the famous Seagrave sedan-body pumpers that Detroit, MI, bought between 1936 and 1967. The trio of revolving lights had changed from the drum shape of the 1920s and early 1930s, to a cone shape.

Fire departments in cities, towns, and villages all over America, had been patching-up 20 to 30-year-old fire engines through 10 years of the Great Depression. They would now have to continue patching up antiquated fire engines through five years of war, plus two to three years of retooling before production resumed.

CHAPTER SIX

Postwar Fire Equipment 1946-1970

The author bought this remarkably well preserved 1950s Fire Grenade demo kit from a firefighter about to retire in Marysville, CA, in 1992. The salesman would hang the glass fire grenade, filled with carbon tetrachloride, over a can of Sterno, and ignite the can. This simulated a fire at a chemical lab. When the heat reached the Fire Grenade, it released its fire-extinguishing chemical onto the can of Sterno. The audience would, presumably, be suitably impressed and order these grenades for their lab: the demo kit included order forms, and envelopes to mail the orders back to the factory. The kit even included a catalog of the various types of fire grenades offered, and a pamphlet the salesman could use to calculate his commission. *Ed Hass*

One of the problems that the fire service had long tried to solve, was how to get water to the upper floors of the ever-taller buildings that architects were designing. New York's Woolworth Building, completed in 1913, had given firefighters a new challenge, until on June 1, 1917, an Ahrens-Fox piston pumper delivered enough pressure through the building's standpipe to produce an effective fire stream 796ft above

Broadway. In 1934, an even taller structure, the Empire State Building, was completed, and it had its own built-in pumps on several floors to force water higher. When a military plane lost in fog smashed into the Empire State Building in 1945, another Ahrens-Fox piston pumper supplied the initial pressure into the standpipe, and the building's own built-in pumps boosted the pressure to extinguish the fire.

Water towers were great for delivering streams as high as to the 8th floor of a building. But they served only that one function. In the Great Depression, it was hard to justify purchasing and maintaining a piece of apparatus that was called out only on those rare occasions when a fire broke out between the 4th and 8th floors of a building. Through the 1930s, demand for new water towers dwindled, and Los Angeles placed the last new water tower in service in 1938. By the time fire apparatus production geared up after World War II, the average water tower had served half a century, its horses replaced by a series of ever newer and larger tractors, and its trailers modified several times to increase its water-throwing capacity. Most water towers were long overdue for replacement.

The solution that had long been sought, was combining an aerial ladder with a water tower as a single piece of apparatus. An aerial ladder could be used to transport men and equipment as high as the 9th floor of a building, and to safely carry trapped fire vic-

Fire extinguishers at New York City Fire Museum include, left to right: double and single cylinder pump handle style, early backpack pump style for small brush fires, and the standard copper soda-and-acid style. After World War II, more sophisticated "Indian" backpack extinguishers appeared, along with new extinguisher types such as pressurized water, foam, and CO_2. By the 1960s, the soda and acid extinguisher was obsolete.

On June 16, 1966, two oil tankers, the *Alva Cape* and *Texaco Massachusetts,* collided in the Kill Van Kull off Staten Island. The spectacular oil fire that resulted killed 30 crew members on the two tankers, and injured 75 more. Fire Boats 7 and 8 of the New York Fire Department, assisted by crews from land-based engine companies and the crews of both tankers, fought the fire from close in. At least two firefighters received medals for rescuing members of the tanker crews. On June 28, 1966, the grateful crew of the Alva Cape presented a section of the tanker's damaged hull plate to F.D.N.Y. This hull plate (right) is now displayed at the New York City Fire Museum, along with the deck monitor (nozzle) of one of the fireboats that did yeoman duty that day (below right).

tims back to the ground. The trouble was that, being made of wood, aerial ladders did not have the strength to double as a water tower. Most wood aerials would have snapped if subjected to the tremendous water pressure that a water tower handled.

The trussed Seagrave aerial had been a good start on the path to strength and rigidity, and some fire departments strung hoses to the top to deliver stream at moderate pressures. But even a Seagrave could not withstand the very high pressures of water tower service.

In 1923, Ahrens-Fox began building air ladders with the Dahill air hoist, as part of their regular product line. They had built a few Dahill-hoist aerials as early as 1916, but only as custom orders. The air hoist was faster to raise and lower, and more stable once raised, than spring or water raised aerials.

In 1930, Ahrens-Fox redesigned their Dahill-hoist aerials, added steel reinforcements sunder the main ladder's side beams, steel hand railings along both sides of the ladder, steel cross-braces on every rung of the ladder, and four steel rods running from the turntable to the top corners of the main (bed) ladder. All of this steel bracing gave the wood ladder enough strength to double as a water tower. Ahrens-Fox installed a high-pressure nozzle at the top of the fly ladder of their new "tower aerial" and sold the first three of these to Newark, NJ, in 1930. Over the next decade, Ahrens-Fox would build 20 more of these water tower/aerial ladder combinations. The tower aerial could reach a height of 85ft, while water towers reached only 55 to 75f high. And a water tower could not carry firefighters and their equipment to the fire floor, or carry fire victims to the ground.

In 1935, Seagrave of Columbus, OH, developed an all-steel aerial ladder, which had even greater strength for water tower service than the steel-reinforced Ahrens-Fox wood aerial did This aerial ladder had a heavy-duty deluge nozzle at its top. It did not require steel rods to brace the top of the bed ladder to the turntable, as Ahrens-Fox used, because the steel ladder itself could support the weight of the tremendous water pressure. The only drawback was that the new Seagrave aerial could only reach a height of 65ft, less than the tallest water

Gamewell of Newton, MA, so completely dominated the U.S. fire alarm telegraph market, that even this stock Fire Alarm Operator uniform patch of the 1960s featured the Gamewell logo, a fist holding lighting bolts. *Ed Hass*

tower. Still, in smaller communities, which had never been able to afford a water tower, the 65-foot height limit of the Seagrave water tower/aerial ladder combination was more than sufficient.

Also in 1935, Peter Pirsch & Sons of Kenosha, WI, developed the first aerial lad-

Safety experts have debated for years whether automobiles or airplanes are safer. Auto-accidents are more frequent, but plane crashes are more spectacular and more deadly. In the 1940s, autos often carried the Pyrene type fire extinguisher at left, while planes were equipped with the slightly larger A-20 aircraft style at right. The author acquired both 1940s vintage extinguishers from firefighter/collector Al Mazzerolle of Marysville, CA. *Ed Hass*

Right: English fire-fighters who belonged to the Emergency Preparedness Scheme (EPS) during World War II received these sturdy military-style helmets. They did not burn and kept heat away from the fire-fighter's skin.

der truck that could not only replace, but even surpass, a water tower. It used hydraulic pumps to force pressurized oil through hydraulic hoses to raise and lower a 100ft, aluminum-alloy ladder. The aluminum was strong enough to support the weight of the water pressure required of water tower service, but lightweight enough not to overwhelm the hydraulic system, even when raised to its full 100ft and throwing water at 100lb pressure. Now, firefighters could rescue citizens from an 11th floor ledge, and pour high-pressure streams through a 12th floor window.

American-LaFrance of Elmira, NY, developed the first all-steel, hydraulically raised aerial ladder in 1938. Like the aluminum Pirsch three years earlier, the four-section ALF aerial reached a height of 100ft. The steel aerial was stronger and sturdier than the aluminum type, but its extra weight placed more demand on the hydraulic hoist mechanism. Annapolis, MD, bought the first

Below: New York City Fire Museum display of ladders, axes, and life nets used during the war years.

one in 1938. Newark, NJ, which had purchased the world's first combination aerial ladder and water tower from Ahrens-Fox in 1930, bought the second of these ALF aerial adder/water tower combinations, also in 1938.

Being the middle of the Great Depression, cities, towns, and villages simply did not have the budget to buy new aerial ladder trucks, even if they did combine the functions of a water tower and a ladder truck on one chassis. So the new breed of aerial ladder trucks from Seagrave, Pirsch, and American-LaFrance did not really sell. With America's entry into World War II, new fire engine production ceased. Even when the war ended, it took most fire engine manufacturers as much as two years to retool and resume fire engine production.

By 1947, nearly all water towers had served for 25 years, and many for 50 years. Many wood aerials had also served two to six decades. Cities all over America began buying the new all-steel, hydraulically-raised aerial ladders in large quantities. Both water towers and wood aerials started vanishing from America's firehouses. By the 1960s, only two die-hard cities, Boston and San Francisco, still maintained their aging fleets of water towers. By the 1970s, only one water tower remained in reserve service in the entire country, a heavily-modified 1897 machine at Memphis, TN.

In 1947, American-LaFrance introduced a revolutionary new fire engine design, the cab-forward, known as their Series 700. These were available as pumpers, heavy rescue trucks, and even hydraulic steel aerials. Placing the cab ahead of the engine, with no hood up front, gave the driver greater visibility. The 700 and 800-series, and the later 900 and 1000 series with dual headlights, would become America's most popular fire engine.

In a brilliant marketing ploy, American-LaFrance had Texaco gas stations give away toys of American-LaFrance pumpers, and induced breakfast cereal manufacturers to include little plastic toy American-LaFrance pumpers and aerials in cereal boxes. Soon, every child in America, and their parents, knew that a fire engine meant American-LaFrance, and their products soon dominated not only the American fire service, but also the American psyche. They were prob-

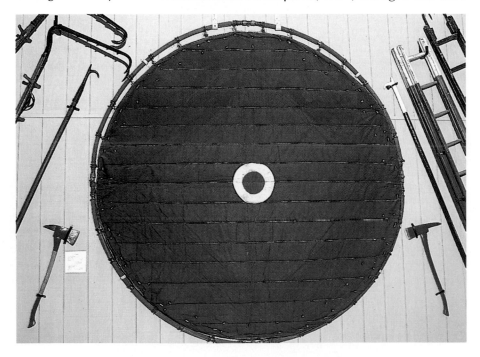

ably the only fire apparatus manufacturer to take advantage of the new postwar industries of advertising, public relations, and product marketing, whose strategy was summed up in the phrase "don't sell the steak, sell the sizzle." Thanks to the efforts of the new Public Relations field, Americans soon knew that when they were thirsty, they should drink Coca-Cola. When they overate they should "be Alka-Seltzer wise." When they wanted to smoke, "Winston tastes good, like a cigarette should." And when their homes or businesses caught fire, an American-LaFrance fire engine would put out the flames.

Thanks to the power of the new Public Relations field, it was difficult for other manufacturers to compete against American-LaFrance's name-brand recognition. Many fell by the wayside. Ahrens-Fox, on rocky financial underpinnings all through the 1940s and 1950s, ceased production in 1958. Only the restyling of the Ahrens-Fox pumper into the Mack C series of 1958-969, and the financial strength of the parent Mack Truck Company, kept Mack fire engine production going. Seagrave faced some rough years before selling out to FWD Trucks of Clintonville, WI, in 1965.

Smaller builders, such as General-Detroit and Buffalo, vanished without a trace. Other small builders, such as Oren (VA), Pirsch (WI), Hahn (PA), Maxim (MA), Ward LaFrance (NY), and Crown (CA), survived by keeping production volume low, and concentrating on regional rather than competing with American-LaFrance for national markets.

Besides the new combination aerial ladder/water towers, the postwar years brought a host of new tools into the firefighter's arsenal.

Before the war, John Bean Company developed a small pump that could spray a stream of high-pressure fog to water orchards. After the war, Bean adapted this pump for firefighting service. Mounted on commercial truck chassis such as Ford or International Harvester, a John Bean pumper looked liked any other conventional pumper, but it delivered a foggy mist of water mixed with firefighting chemicals, over a wider area than conventional straight streams of water. High-pressure fog was ideal for fighting the new fire hazards of the

mid-20th century, such as automobile collisions, chemical spills, and airplane crashes.

Speaking of airplanes, they had been pretty much a novelty before the war, used in barnstorming and flying circus shows, or for freight service such as carrying mail between major cities. After the war, airlines such as TWA, Pan Am, American, and United began offering regular passenger service, with established routes between airports in most major American cities. Airplane manufacturers, such as Douglas and Boeing, were constantly developing larger planes, to carry more passengers on each flight. Firefighters now faced a new hazard: these huge airplanes sometimes crashed, spilling and igniting large amounts of fuel, and endangering dozens of the plane's passengers and crew.

American manufacturers responded to this new challenge with a new type of fire engine, the airport crash truck, and a new firefighting chemical agent, foam. Covering spilled aircraft fuel with a blanket of foam, deprived a fire of oxygen. As every Boy

In 1930, Ahrens-Fox built the first aerial ladder truck strong enough to double as a water tower. The first of these steel-reinforced wooden "tower aerials" is seen undergoing a test near the manufacturer's Cincinnati factory, before delivery to Newark, NJ. Cincinnati's two 1917 1,200-gallon-per-minute piston pumpers are at left, supplying the water to the top of this 85ft ladder. *Photo from the late Robert R. Johnson, former Ahrens-Fox salesman*

111

Scout of that era knew, the Fire Triangle consists of heat, fuel, and oxygen. Deprive a fire of any one of these three, and it ceases to burn. The traditional use of cold water on fires deprived a fire of heat. Blowing up buildings or bulldozing trees in the path of a fire deprived it of fuel. And now foam could deprive the fire of oxygen. Walter, Cardox, Oshkosh, and of course American-LaFrance, were the major suppliers of this new airport fire apparatus.

Not only did firefighters have a new type of equipment for fighting airplane fires, but the airplane itself became a new weapon in the firefighter's arsenal after World War II. A tank mounted under the airplane could be filled with water, foam, or other fire extinguishing chemicals, and its contents released above a forest fire. The U.S. Forest Service, California Division of Forestry and Firefighting, and Los Angeles County Fire Department were pioneers in the use of water drops from aircraft, and they rely heavily on this method to extinguish forest fires today.

But water drops alone cannot extinguish wildland fires, and in the late 1940s, a new breed of firefighter appeared: the smokejumper. A prepared bundle of firefighting tools, such as axes, shovels, brooms, and backpack-style fire extinguishers, was connected to parachutes and dropped out of an airplane into the heart of the fire. The "smokejumper" firefighters parachuted out right behind these tools, and set to work putting out the fire on the ground, while airplanes dropped their fire extinguishing chemicals from above.

One other new firefighting tool helped extinguish forest fires. Although developed for quickly tearing down old buildings to make room for new ones, the bulldozer was ideal for quickly clearing trees and underbrush. Bulldozers could clear a fire break wide enough that a raging forest fire could not leap over it. The fire, deprived of fuel, would then die out. The smoke jumpers then sifted and drowned the smoldering ashes, so the fire could not rekindle.

The smoke jumper's backpack fire extinguisher is worthy of note as another new tool for extinguishing forest fires. Developed in the 1930s, and originally made of heavy brass, they were made of lightweight aluminum by the 1950s. The "Indian Tank" by

Thanks to their public relations campaign and high sales volume, the American-LaFrance 700-series cab-forward apparatus became synonymous with fire engines in the 1940s and 1950s. This hydraulic-raise, steel aerial on 700-series chassis served Fremont, CA, and now belongs to the Fire Department Muster Team of San Jose, CA. *Ed Hass*

Above: Maxim Motors of Middleboro, MA, built a great many fire engines between 1915 and 1980, but few ever made it to the West Coast. This superbly restored exception, lettered for Red Hills Ranch, appeared at the Firefighters Muster in Columbia, CA. *Ed Hass*

Right: One look at this tough searchlight truck from F.D.N.Y. and it is no wonder that Mack "B" models are so poular with antique fire truck collectors. After 31 years on the front line, this workhorse has earned its rest in the New York City Fire Museum.

Left: Fire Dept. of Kings County, CA, stationed this circa 1965 Kaiser Jeep at the fire station in Kettleman City, CA. The tough 4x4 carried water, hose, and tools to rural grass and brush fires. *Ed Hass*

Below: One of a fleet of 25-year-old retired fire engines parked outside and awaiting disposal at San Jose Fire Dept. Repair Shop on Bird Ave. This FWD pumper was built in Clintonville, WI.

FIRE EQUIPMENT

D.B. Smith Company of Utica, NY, was the most popular of these backpack extinguishers. Two shoulder straps allowed the firefighter to carry the Indian Tank into the heart of a forest fire. A long rubber hose, connected to the bottom of the extinguisher, ended in a long nozzle. Rapidly sliding the outer section of the nozzle back and forth over the inner section, sprayed the extinguisher's contents directly onto the burning vegetation, whether it was trees, bushes, or grass.

The era of the Great Fires, which destroyed entire cities, had ended with the Great Baltimore Fire of 1904 and the San Francisco Earthquake & Fire of 1906.

Although not nearly as widespread as these Great Fires, the 1928 Pocasset Mills fire in Fall River, MA, had wiped out many city blocks. At the Chicago Stockyard in 1934, and again at the Rosenbaum Grain Elevators in 1939, Chicago had required 100 of the city's 129 pumpers.

In 1947, America would be shocked out of its false sense of security by one more Great Fire that destroyed an entire community, this time at Texas City, Texas.

Located across Galveston Bay from the city of Galveston, Texas City was just 11 miles from the Gulf of Mexico. During World War II, Monsanto and other chemical companies had located there, to supply nitrate explosives and other materials for the war effort. Oil from Texas wells was refined in Texas City, too, and the city had also become an important hub for rail freight from all over the South. With a harbor 800ft wide and over five miles long, ships from this busy port had transported men and materials to war zones all over Europe. Over $125 million had been invested in the city's industries, and 16,000 workers had been added to the city's permanent population.

Texas City had a 47-man fire department, entirely volunteer except for the paid drivers of the city's four fire engines. Shortly after 8:00am on April 16, 1947, 27 of these firefighters were called to a report of fire aboard the Grand Camp, a freighter had docked at Texas City five days earlier, taking on a 2,300-ton cargo of ammonium nitrate, used in making artificial fertilizer. Because this nitrate was in granulated form, this already-explosive chemical had been laced with highly-flammable petroleum, to prevent caking.

Because the Monsanto plant was in process of a $1 million remodeling and expansion to switch from making nitrate explosives to polystyrene plastics, the ship's cargo had been shipped to Texas City by rail from chemical plants in Iowa and Nebraska, rather than produced locally. This partly explains why the ship had been in Texas City for five days to take on cargo.

Across the same slip, at Pier B, the freighter Wilson B. Keene had taken on a cargo of 961 tons of this same ammonium nitrate.

On April 15, several crew members of the Grand Camp had been observed smoking cigarettes while on a break from loading the cargo. Perhaps one of them accidentally flicked an ash through an open cargo door. Or perhaps one deliberately tossed his cigarette into the cargo hold so the ship's officers would not see him. At any rate, a cigarette somehow got into the cargo before the hatch doors were closed at 5:00pm. When the doors reopened at 8:00am on the 16th, smoke from the fire that had been smoldering all night poured out of the cargo hold. That is when the Texas City firefighters were summoned.

Aided by the Grand Camp crew and dock workers, the Texas City volunteer firemen stretched hoses from their four pumpers to the deck of the freighter. About 8:30am, fire streams began pouring into the Number 4 Hold, where the fire was first spotted.

The Texas City firefighters had not been informed that the cargo was ammonium nitrate, and even if they had been told, they were probably unaware that similar nitrate compounds were used to make explosives.

At 9:12am, the overheated cargo exploded, with a force strong enough to lift a nearby 75ft steel barge and toss it on shore. The fire instantly spread to the freighter Wilson B. Keene, docked nearby, and to the Monsanto chemical plant and the oil refineries. The shock wave of the blast knocked loose the plastic processing equipment inside the Monsanto plant, suddenly releasing a pressure of 1,000psi, more than ten times the pressure required to knock a hole through a human body.

A blast-induced fire aboard the freighter High Flyer, also loaded with ammonium nitrate, heated the fertilizer enough to cause this ship to also explode, at 1:10am on April 17. This second blast ignited three more nearby oil refineries. One refinery alone contained 80,00 barrels of oil.

The fires and blasts not only leveled most of the town's buildings, but killed 433, and 302 more were missing and never found in the wreckage. Several thousand more were injured. Property damage was estimated between $50 and 125 million, including 500 railway cars and 1,500 motor vehicles. More than half of the city's firefighters died in the blast, and all four Texas City fire engines were destroyed.

Lessons of this fire included:

Left: Not a brilliant gadget! The Hero "fire extinguisher in a can" was released when a tab was pulled—much as a soda can is opened today. The resulting release of carbon-tetrachloride was insufficient to put out all but the smallest fire.

Below and Bottom left: Fog nozzles found popularity in the 1950s, allowing firefighters to control the water stream from a solid, straight blast to a misty fog. These nozzles fit anything from the standard 2.5in supply-line hose (bottom) to the 1in booster hose for smaller fires.

• Don't concentrate several high-hazard industries, such as chemical plants, oil refineries, and cargo ships, in one location.

• When planning how to handle a disaster, consider not only the danger at the source of a fire or explosion, but the exposure risk of other nearby buildings.

• Firefighter training must include knowledge of new chemical hazards and how to deal with them.

• All chemicals that have a high danger of fire or explosion must be clearly labeled as to their composition and dangers.

• Transporting and loading of dangerous substances must be more strictly regulated, controlled, and supervised, especially enforcement of No Smoking regulations.

• Communities should have mutual-aid agreements with nearby communities, in case a disaster proves too great for the local fire department to handle alone.

L&G of Liverpool, England, made this fire blanket during World War II. It was designed for personal use.

The postwar era would also revolutionize the motive power, and the pump type and capacity, of fire engines. This would be the fourth major change in the way fires were fought in America. The first occurred in the 18th century, with the switch from buckets to hand pumpers. The second was the post Civil War switch from hand pumpers to horse-drawn steam fire engines, aerial ladder trucks, and water towers. The third was the early 20th century shift from steam and horses, to gasoline motors powering piston, rotary, and centrifugal pumpers.

The years after World War II would bring a shift from gasoline to diesel engines, abandonment of piton and rotary pumps, use of automatic transmissions, and increased capacity of centrifugal pumps from 750 or 1,000 gallons per minute (GPM), to 1500 or 2000 GPM.

New Stutz Fire Engine Company of Hartford City, IN, delivered America's first diesel fire engine to Columbus, IN, in 1939. Chicago Fire Dept. installed a Cummins diesel engine in an aging Ahrens-Fox pumper in 1943, and another diesel in a fireboat that same year.

Big-capacity pumps had always been a rarity. Even in the steamer era, a very few piston pumps had been built with capacities of 1,000 to 1,3000 gallons per minute (GPM). Ahrens-Fox had a 1,300 GPM pump as part of its standard product line as early as 1919, and built two 1,500 GPM pumpers for New York in 1933. The Crown Firecoach of Los Angeles, introduced in 1951, was the first to make large capacity pumps a standard part of its product line, building its first 1500 GPM pumper for Los Angeles in 1954, and the world's first 2,000 GPM pumper the following year.

The first diesel-powered Crown, a 2,000 GPM pumper for Vernon, CA, appeared in 1964; its 855 cubic inch Cummins produced 380 horsepower.

Mack Trucks built its first diesel fire engine, powered by Mack's own Thermodyne diesel engine, in 1959. The following year, Mack sold three diesel pumpers to Hamilton, Bermuda. Mack had already started building fire engines with automatic transmissions in 1957.

In 1960, American-LaFrance of Elmira, NY, briefly tried building fire engines powered by jet turbine engines. They sold the

1960 Cairns "New Yorker" leather fire helmet.

first, a 100ft tillered aerial ladder truck, to Seattle, WA, then built turbo jet pumpers for San Francisco, CA, and Mount Vernon, VA. The jet engines were extremely noisy, and the brakes were not adequate for the engine power. Within a few years, all three were converted to gasoline engines.

In 1961, American-LaFrance built its last fire engine with the V-12 motor, based on the Auburn-Lycoming automobile engine of he 1930s. From now on, all American-LaFrance apparatus would be powered either by a Continental gasoline engine, or a diesel engine.

Just as the electric siren of the 1930s had made hand-cranked mechanical sirens obsolete, the new electronic sirens introduced in the late 1950s and early 1960s, made electric sirens obsolete.

The days of double-clutching every gear, whether sifting up or down, were drawing to a close. Standard transmissions now had synchro-mesh gears, so fire apparatus drivers no longer had to shift into neutral before shifting another gear. Better yet, more and more manufacturers were coupling the new breed of diesel engines to automatic transmissions, which seemed a natural marriage of two improved mechanical components.

In 1957, Robert Quinn, Fire Marshal of the Chicago Fire Department, strung a fire hose along the elbow-like mechanical boom of a street light repair truck, and an entirely new breed of fire engine was born. The elevating platform, or Snorkel, could bend around overhead telephone wires, and its bucket-like platform could carry men and equipment up to a fire floor, and transport rescued fire victims back to the ground, more quickly, efficiently, and safely than aerial ladder trucks. And the nozzle at the end of the basket let it double as a water tower.

In 1958, two new fire engine manufacturers spring up to build elevating platforms: Pitman Manufacturing Company (later renamed Snorkel Fire Equipment Company) in Missouri, and Hi-Ranger in Fort Wayne, IN. Instead of stringing rubber fire hose up to the basket, as Quinn had done on the 1957 prototype, these elevating platforms used steel tubes to bring water up to the built-in deluge nozzle at the top.

The Pitman Snorkel used time-tested hydraulics, similar to those on aerial ladder trucks, to raise and lower the elevating platform. The Hi-Ranger, however, used steel cables wound and unwound on a drum. After a few accidents involving snapped cables, the Hi-Ranger brand quietly disappeared from the market.

In 1962, American-LaFrance came out with its own elevating platform, called the Aero-Chief. When one of these overturned from improperly-set outriggers, killing three Syracuse firefighters in the late 1960s, the Aero-Chief also lost popularity. Although the accident was the fault of over-eager firefighters who had raised the elevating platform before the supporting outriggers were in place, the Aero-Chief's reputation was ruined. Its production ceased by the mid-1970s.

ATO, which owned American-LaFrance, then acquired Snorkel ire Equipment Company, and began offering Snorkel's popular elevating platforms on American-LaFrance chassis, as ell as on commercial truck chassis such as Ford

In 1970, one of Snorkel Company's elevating platforms also overturned from improperly se outriggers, at Los Angeles, CA. But Snorkel Company promptly responded by including safety locks on all of their new apparatus from then on, and installing these locks on all Snorkels already in service. An elevating platform could no longer be raised, until the outriggers were fully extended and locked in position. These outriggers prevented the fire engine from tipping over while the Snorkel was raised.

In 1965, Mack Trucks trumped everyone in the race to gain ever more capacity from their centrifugal pumps, building a tractor-trailer pumper for New York that could pump 8,800 gallons per minute at 350lb

pressure. Dubbed the "Super Pumper," it coupled Mack's diesel-powered Model F tractor to a custom trailer carrying a DeLaval marine pump powered by a huge Napier diesel engine. Gibbs & Cox of Philadelphia, famous for its gigantic and powerful fireboats designed the apparatus. In fact, the Super Pumper was publicized as a land-based fireboat.

The 8,880 GPM capacity could not be reached using conventional 4-6in hard suction hoses, so the Super Pumper used two hard suctions with an incredible 10in diameter. The back of the pumper was fitted with a crane boom to raise and lower these massive suction sleeves. Where a river or the ocean was not near the fire scene, the Super Pumper drew water from not one, but four to eight separate fire hydrants. A second tractor-trailer Mack Model F unit, the "Super Tender" carried the pumper's 1,500ft of 3in discharge hose.

The pump capacity of the Super Pumper would not be exceeded on a land-based fire pump for nearly 30 years, until Rotterdam, Holland, bought a fixed-location, 20,000 GPM fire pump in 1993. Since that pump is not mounted on a vehicle and is not portable, it is not really a fire engine in the conventional sense. So the super pumper is still the reigning champion in pump capacity for any fire engine.

In August, 1969, the first astronauts walked on the moon. An incredible amount of new technology, especially new types of computers, was developed to make that first moon walk possible. It would not be long before these new technologies would be applied to all aspects of everyday American life, including the fire service.

As the 1970s dawned, fire departments would use computers and sophisticated electronics for such diverse applications as dispatching fire engines and maintaining two-way radio communications at a fire, to controlling fire pumps. And the new light-weight materials developed for use in weight-sensitive spacecraft would show up in everything from uniforms and turnout coats, to fire hose and hose couplings. As we shall see in the next chapter, the fifth revolution in American fire service history, high technology, would sweep the fire service in the 1970s and 1980s, just as it did the rest of American life of that time.

Fisher Beer wasn't a brewery, but a 5 and 10 cent store chain. When their store in Hempstead, NY, burned in the 1940s, two of the town's fleet of Ahrens-Fox apparatus turned out. In front is 1927 Model M-S-2 piston pumper #1742 of Union Engine Company #2. Behind it, at left, is 1927 Model 85-6-1 #2035 85ft tillered aerial, of Harper Hook & Ladder #1. Courtesy Herb Barber

CHAPTER SEVEN

Fire Equipment of Today 1970-2000

The economic slump of the late 1970s and early 1980s saw a long list of established fire engine manufacturers vanish.

Oren of Roanoke, VA, was absorbed into Grumman (of military aircraft fame), then into Food Machinery Corporation (FMC), and from there into corporate obscurity.

Mack Trucks was absorbed into French truck maker Renault, which in turn became part of the Chrysler-Plymouth-Dodge-Jeep-Eagle conglomerate. Mack truck chassis would still be available for others to mount fire engine bodies on, but Mack stopped making complete fire engines.

Maxim Motors of Middleboro, MA; Ward LaFrance of Elmira Heights, NY; and Sanford of Syracuse, NY, all ceased to exist about the same time, 1980 or 1981. The author knew a fire apparatus salesman who had the misfortune to represent all three companies, and to lose all three jobs to corporate demise within the same month.

Peter Pirsch and Sons of Kenosha, WI, pioneers in design of metal aerials and hydraulic hoists, also folded, as did the west coast's two favorite fire engine brands, Crown Firecoach of Los Angeles and P.E. Van Pelt of Oakdale, CA.

American LaFrance, which had somewhat lost its corporate identity when A-T-O acquired it in the early 1970s, was sold to Figge International, a corporate giant that owned a wide variety of industries. Although many of American-LaFrance's sales in the lean years of the Great Depression were due to local unions supporting purchase of union-made ALF apparatus, new owner Harry Figge despised labor unions. He closed the Elmira plant, and reopened as a non-union shop in Virginia.

Most firefighters belong to the International Association of Fire Fighters (IAFF) union. They did not forgive or forget Figge's union busting. American LaFrance, once outselling all other brands combined, saw its sales slump drastically. Within a few years, American LaFrance was no more. In 1995, Freightliner Trucks acquired what was left of the once-proud manufacturer, whose roots go back to 1832. Today, Freightliner offers its own truck chassis, with Waterous pump, under the American LaFrance brand name.

Part of the reason so many fire engine manufacturers folded at the same time was the general lack of economic health in American industry at the time. Traditional "smokestack" industries, such as steel and automobiles, were closing their plants and laying off thousands of worker.

Another reason was the time needed to build a fire engine with all of the new high technology features. It could take up to a year from the time the fire engine manufacturer began negotiating a sale, drawing up specifications, and submitting a bid proposal, until the contract was actually signed. To be the successful bidder, the quoted sale price had to be lower than any competitor's.

Two firefighters unload hose from a 1985 Crown pumper at San Jose, CA, to transfer it to a newer pumper from Hi-Tech of Oakdale, CA. But it is not just the brand name of the new pumper that is high-tech. The hose is of a lightweight but very tough and flexible plastic, eliminating the need for the mildew-prone woven cotton jackets of traditional rubber fire hose. The yellow "bunker pants" are treated with Nomex, a new chemical that is highly resistant to flames and extreme heat. The nozzle is made of a new space-age metal alloy that is twice as strong as, but less than half the weight of, traditional brass hose couplings. A 50ft length of traditional cotton-jacketed hose with brass couplings weighs 50lb, while the new hose weighs about 20lb.

Another view of the hose unloading from a 1985 Crown pumper at San Jose, CA, to transfer it to a newer pumper from Hi-Tech of Oakdale, CA.

1985 Crown pumper at San Jose, CA.

Pump panel of 1985 Crown pumper at San Jose, CA, shows the beginnings of new space-age technologies. Unlike traditional steel pump panels, the polished aluminum pump panel did not rust from water dripping out of the pump, and it was easier to keep clean than a painted surface was. The ABC fire extinguisher on the running board could handle any type of small fire, from grease burning on a stove, to a short in household electric wiring. The trigger handle on the extinguisher was easy to grip and squeeze, to spray the fire-extinguishing chemical on the fire. The plastic knobs on the pump valve shutoff handles were tough, durable, and lightweight, with easy-to-grip knurled edges and palm-fitting ball shape. The pumper itself featured such modern amenities as 2,000 GPM pump, diesel engine, automatic transmission, and fully-enclosed cab.

San Jose, CA, is America's eighth largest city and California's third largest. Center of much of the world's computer and electronics - in "Silicon Valley" - it still contains many woodlands on steep hills. To cope with fires here, San Jose developed the "brush patrol" on a four-wheel-drive pickup truck.

Cab-door lettering shows the technique of swirling patterns into gold leaf letters, introduced in the late 1950s. In days gone by, this type of gold leafing involved sketching the lettering onto a paper pattern, laboriously punching holes into the pattern, and tapping the pattern onto the door. The gold-leafer then pounded graphite dust through the pattern holes, removed the pattern, and applied a special glue called gold sizing to the areas he would gold-leaf. He would then press squares of gold leaf onto the door, let the glue dry, and brush away the gold that did not stick, leaving gold only in the shape of the letters. The artists then twirled a round file into each gold letter to get the swirling effect. Today, gold leaf artists can type the word into a computer, and print out a strip of properly-spaced and pre-swirled gold-leaf letters, ready to glue on. In restoring his own 1953 fire engine, the author watched a gold-leaf artist spend a full day to letter two cab doors using the old technique, and then produce equivalent results on both sides of the hood in about one hour, using computer-generated gold leaf lettering strips.

In the 1960s, civil unrest in big cities meant that anyone in any type of uniform was likely to be shot at, even firefighters. So rear steps became too small to ride on, and cabs became fully enclosed, including the previously-exposed jump seats at the rear of a cab. Cabs also grew large enough for five to seven firefighters. By the 1980s, seats in these new cabs became plushly padded, with waist and shoulder seat belts. In addition to diesel engines and automatic transmissions, the new high-technology fire engines featured power steering, air brakes, electronic sirens, and even air conditioning! This example of the new high-technology style of space-age fire engine, built by the aptly named Hi-Tech company of Oakdale, CA, serves as Engine 26 in San Jose, CA.

The technology in the new fire engines was too complex for one manufacturer to make everything themselves, as they had in the 1920s. So pumps were bought from Hale (Conshohocken, PA) or Waterous (St. Paul, MN), truck chassis from Spartan, Hendrickson, or Warner & Swasey, cabs from Truck Cab Manufacturing of Cincinnati, special firefighters seats from 9-1-1, and so on. The manufacturer would assemble these components, and make their own steel fire engine bodies around it all. With low-cost American steel no longer plentiful due to steel mill closings, fire engine manufacturers had to pay the higher cost of importing steel from Japanese steel plants.

Obtaining the components to make a complete fire engine could take a long time. For example, Allison automatic transmissions were too expensive for manufacturers to keep a ready stock against sales that the might not make. So the transmissions had to be ordered as needed, and it could take up to a year to receive one from Allison.

Then it took a year or more to assemble the complete fire engine, which was a far more complex machine than the fire engines built in 60 to 90 days in the 1920s. By the time a new fire engine was delivered,

3 or 4 years had elapsed since the salesman's original contact with city officials. The manufacturer still had to sell the fire engine to the town at the agreed-on price, without adjusting for 3 to 4 years of monetary inflation. It was hard to stay profitable under these conditions.

At least four new fire engine manufacturers stepped up to fill all of these vacancies in the fire engine industry.

Smeal, in Nebraska, had been a successful builder of commercial truck bodies before launching its own line of pumpers, and aerial ladder trucks based on the company's successful construction crane technology. Like many manufacturers, their first few fire engines were subject to quality problems. The author recalls one of the first of their new aerial ladder trucks undergoing repairs to its front springs before it could be placed in service). But today, Smeal is known for its quality products.

Pierce, of Appleton, WI, had been making fire apparatus bodies, especially for Snorkel fire engines, before launching its complete line of pumpers, aerial ladder trucks, and rescue trucks. Today, many consider Pierce equal to or above Seagrave for the finest quality fire engines on the market.

Canadian Fire Engine builder Pierre

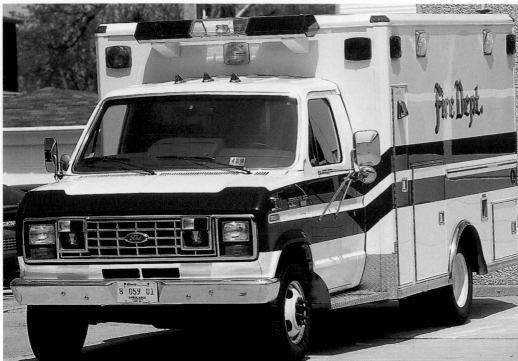

Above: With today's fire departments handling far more medical calls than fires, many run their own paramedic vehicles and even ambulances. While these huge, boxy ambulances on Ford, Chevy, or GMC truck chassis lack the comfort of traditional Cadillac-based station-wagon ambulances, they carry far more medical equipment.

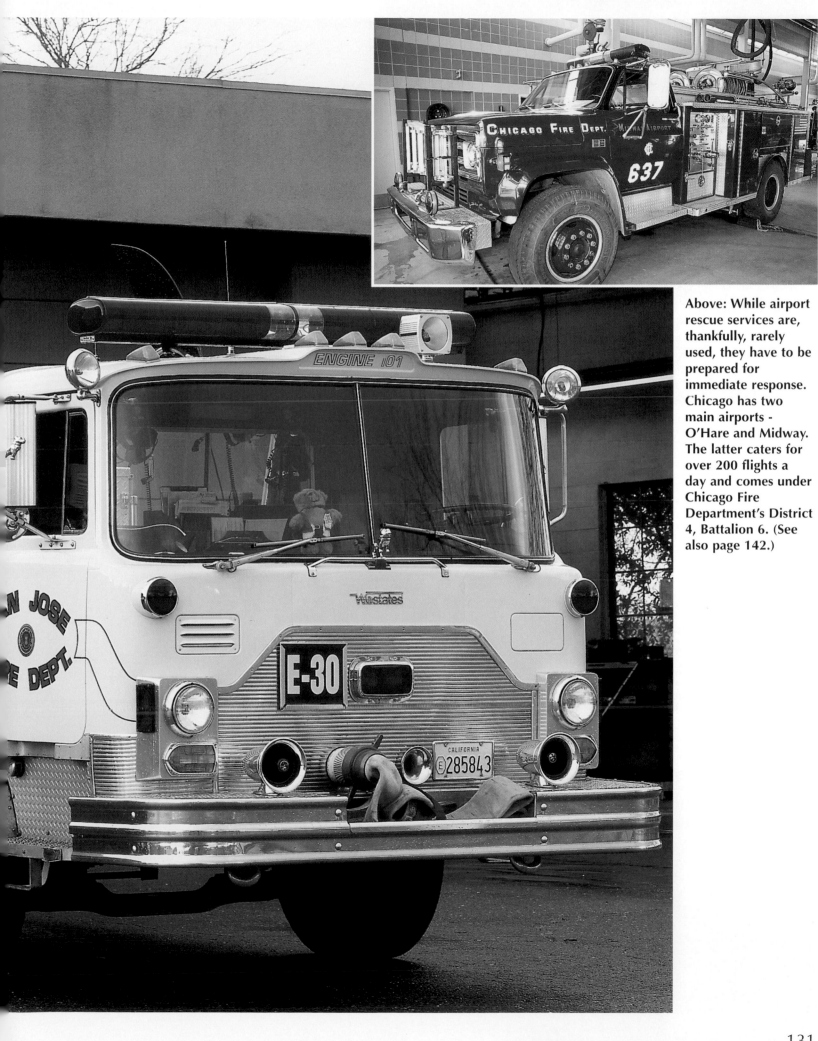

Above: While airport rescue services are, thankfully, rarely used, they have to be prepared for immediate response. Chicago has two main airports - O'Hare and Midway. The latter caters for over 200 flights a day and comes under Chicago Fire Department's District 4, Battalion 6. (See also page 142.)

Thibault, renowned for its nearly-indestructible aerial ladders, was absorbed into Nova-Quintech. In 1997, Pierce acquired Nova-Quintech, giving customers a choice of Pierce's own aerial ladder design, or the large and sturdy Thibault style.

Hi-Tech opened in Oakdale, CA, with a brand new factory using the latest high-technology manufacturing processes. In 1996, I watched as workers at Hi-Tech used the latest welding technologies, to build a 500-gallon booster tank for a new pumper. Being in the same city where Van Pelt had built fire engines since 1925, Hi-Tech hired many of the laid-off Van Pelt employees for their knowledge and experience.

Emergency-One, often called E-One for short, began fire engine production in Florida with a new strategy. They offered only one standard model of pumper, and one standard ladder truck, and pre-assembled as many standard components as possible. The customer had to accept the standard model, with little room for deviation. When E-One made a sale, they would use these pre-fabricated components, to complete the fire engine as quickly and inexpensive as possible. Fire Chiefs, many of whom like to think that the firefighting requirements of their community are unique, did not like having to accept the manufacturer's specifications, instead of the manufacturer accommodating the chief's specifications. Municipal accountants loved the cost savings, and having the apparatus delivered in less time.

In the new order of things, Pierce would dominate the high-end fire engine market, the lofty "Rolls-Royce" heights occupied by Ahrens-Fox in the 1920s, and by Mack in the 1960s and 1970s, with carefully crafted custom-built fire engines that performed exceptionally well even under the toughest conditions.

Seagrave would hold its traditional middle ground, a high-quality 'Cadillac' of custom fire engines at reasonable prices. E-One and Smeal took over American LaFrance's control of the 'Chevrolet' level low-end of the market, the tough work-horse that underbid all competitors, sold more than any other make, and could meet most normal firefighting conditions.

About 1977, the author's hometown, Plainfield, NJ, was one of the first to buy a

new pumper with a computer control. Computers had only recently changed from occupying an entire room, to fitting on a desktop (Apple). But this computer sometimes engaged the fire pump while the fire engine was driving, causing the rear axle to suddenly lock up. Today, computer controls on pumps are common and work very well to monitor water flow. For example, if a firefighter inside a burning building shuts off a hose line too quickly, the computer automatically adjusts for this, preventing a destructive "water hammer" inside the pump.

Computers have moved into the cab, too, showing the driver the best route to the fire, and showing the officer at right front of the cab, what hazardous materials might be in the building, and how to handle them. For example, it is helpful to know that a building stores magnesium, which explodes when wet; or paint thinner, which not only explodes, but when heated produces fumes that are a known carcinogenic (cancer-causing) agent.

Personal computers inside the firehouse are used for everything. They log reports of

Left: A modern firefighter's helmet, as used by the Captain of Engine 26 in San Jose, CA, is made of tough space-age plastic. It can protect the wearer's head from the impact of nearly any weight of falling objects. The clear plastic face shield is resistant to cracking, shattering, and scratching, even under extreme interior firefighting conditions. Note the adjustable chin strap, treated with flame-resistant chemicals.

Far left: A variety of modern fire extinguishers, from water and foam types to the universal ABC style, awaiting refilling at the San Jose Fire Department Repair Shop on Bird Avenue in San Jose, CA. Once recharged, they will be reinstalled on fire engines, or on the walls of city office buildings, and commercial or industrial structures.

FIRE EQUIPMENT

fires, car accidents, rescues, and medical calls. They keep inventory of fire hose and other equipment: when it was purchased, how often it is used, and any signs of wear or damage. Special software is available to aid in firefighter training.

Computers have also taken over the fire alarm dispatch office. They tell the dispatcher the phone number and address from which a call originates, and the cross street nearest to that address. The computers also show what fire companies are currently available (not out on other calls) in the immediate area, so the dispatcher can promptly send the right equipment.

In a life-threatening emergency, every second counts. To save time in dispatching firefighters and equipment, fire alarm dispatchers have developed their own shorthand. If you listen to fire calls on a home scanner, you are likely to hear such jargon as IC (Incident Commander, the ranking fire department officer who directs operations at a fire, auto accident, or other emergency), or RP (Reporting Person, the person who reported the emergency).

Or you might hear first due (the fire engine and crew who arrive at the emergency first, size up the situation, and decide what needs to be done immediately).

For fires in grass fields or forests, called wildland fires, you might hear a dispatcher refer to fuel temperature. This does not mean how hot the gasoline is in the fire engine, but the temperature of the burning grass or trees.

You might hear the word "versus," but this does not mean a boxing match (Ali versus Frasier) or a court case (the United States versus 500 irate taxpayers). The meaning of automobile versus pedestrian, or big rig (tractor-trailer truck) versus train, is pretty clear.

You might also hear the word 'structure'. This means the emergency is indoors, in a house, factory, warehouse, office building, or any other type of man-made ñ well, structure.

Sometimes, listening to the fire department dispatcher on a home scanner can lead you to interesting adventures, too. One morning a few years ago, I heard a call from the San Jose Fire Department dispatcher for 'helicopter versus structure' and instantly knew who won that contest. A helicopter

had been positioning a heavy air conditioner onto the roof of one of he city's tallest office buildings, when a sudden gust of wind unbalanced the chopper, sending it and the air conditioner crashing onto the roof, and scattering bits of the helicopter all over the downtown area. Seven alarms were called, including mutual aid (firefighters and fire engines) from the nearby communities of Santa Clara, Sunnyvale, and Mountain View. I watched the efforts to rescue the helicopter pilot (they were too late to save him), and the next day watched Federal Aviation Authority (FAA) officials pick up, wrap, and label every piece of the wrecked helicopter, from the roof, parking lots, and streets. In their extensive lab, each piece would be meticulously examined and tested, to determine the cause of the crash, and how to design safer helicopters that would not suffer a similar fate.

Another exciting dispatch I heard several years ago, started while I was driving home late one evening. Reversing normal procedures, Engine 5 of San Jose, CA, called the dispatcher to announce that they were dispatching themselves across the street, as they had spotted flames and smoke pouring from a printing shop. The quantity of paper and printer's ink, typically found in such a shop, would guarantee plenty of fuel to feed this fire. Moments after they arrived, the Captain of Engine 5 requested a second alarm, so a pumper, aerial ladder truck, rescue truck, and battalion chief were dispatched to the fire.

In the time it took me to get across the city streets of Sunnyvale and head for the highway, to take me the ten miles to San Jose, the Battalion Chief had arrived. He instantly took charge as IC (Incident Commander), and just as I reached the highway, he requested a third alarm, reporting that the entire structure was in flames.

Soon, I was about halfway to the fire, passing through Santa Clara. I could already see the flickering flames and billowing smoke against the night sky, even though I was still 5 miles away.

Soon after, I heard a panicked firefighter's voice: 'Everybody of the roof. NOW!' This firefighter's microphone was still open when I heard the terrific creak and crash of the print shop's roof collapsing into the fire. Fortunately, all the firefighters evacuated the

In the 1970s, nearly every pumper, ladder truck, and heavy-rescue truck in the Fire Department of New York (F.D.N.Y.) was a Mack. F.D.N.Y. is the world's largest municipal fire department, with over 300 engine companies and over 100 ladder companies, so that is quite a quantity of Mack fire engines in one city. Although the city now has some American-LaFrance, Seagrave, and Sutphen fire engines, the vast majority of New York City's fire engines are still Mack. This is the interior of a c.1980 Mack Model CF pumper, still serving Engine 24 in midtown Manhattan. Note the computer console at far right. Air bottle and helmet are strategically placed for a firefighter to put on while en route to a fire. Seats are specially designed to provide firefighters with maximum comfort and safety.

Even with space-age technologies, some aspects of fire engines remain unchanged since Colonial days. This traditional fire axe is mounted on a 1985 Crown pumper at San Jose, CA.

Firefighting equipment carried on F.D.N.Y. Truck 5, a tillered Seagrave aerial, includes reel of electric wire to connect the portable generator to searchlights and other electric tools; acetylene cutting torch; pipe wrenches and gas shutoff wrench; door pry bar; and extra helmets, air bottles, and gas masks. The circular power saw (seen atop the fender of the ladder truck's tractor) ventilates roofs of buildings much more quickly than the traditional fire axe. The Hurst Jaws of Life, at lower right, quickly cuts open the door and roof of a car or truck at the scene of an automobile accident, so paramedics can remove the accident victim and provide emergency medical aid.

Not too many years ago, a firefighter's comfort and safety was not a major concern of the municipal officials who ordered fire equipment, or manufacturers who built it. The author's own 1953 Ahrens-Fox pumper was built without a cab roof, so the town could save the few dollars it cost for the extra metal and labor to include a roof. During summer rains, and winter blizzards, the firefighters aboard this truck were exposed to the elements. In the 1970s, someone figured out that letting firefighters arrive at a blaze wet, cold, and exhausted meant they were not at peak efficiency. The new concern for a firefighter's comfort and safety resulted in fire engines like this tractor-trailer Seagrave aerial ladder truck of F.D.N.Y Truck 5, in midtown Manhattan's Greenwich Village. Air suspension means that arriving firefighters are no longer sore from the jostling of traditional steel spring suspension. Hearing protectors prevent temporary (and sometimes permanent) hearing loss that can result from prolonged exposure to the roar of the diesel engine, the blast of the air horn, and the wail of the electronic siren. Firefighters today arrive in top physical and mental condition for the grueling job awaiting them.

roof before it caved in, so nobody was injured or killed.

Just as I was parking my car to walk over and look at this fire, the IC called for a fifth alarm. I carefully stepped over the main 5-inch plastic supply hose to get my first up-close look at this five-alarm fire. Despite being twice the diameter, this supply-hose actually weighs less than conventional 2-.5 inch cotton-jacketed rubber fire hose.

I positioned myself where I could see all the action, but was still at a safe distance from the fire. There, I would not interfere with firefighting operations, or risk burning or smoke inhalation. And, of course, I stood upwind of the fire, so the billowing, heavy, black and toxic smoke, laden with fumes of printer's ink, blew away from me, not toward me.

I then set up my video camera, getting spectacular video footage as two 100-foot steel aerial ladders, each with two water tower nozzles at the top, pelted the leaping flames and smoke from above. A fleet of 2000-GPM centrifugal pumpers, parked all over the neighborhood, supplied water to the four tower nozzles, and to over a dozen hand-held hose lines.

Firefighters worked through the night, but the fire ended only when the printing shop had burned to the ground. Spraying a curtain of water around the building, San Jose firefighters had prevented the fire's spread to other nearby structures.

Soon after I began taping, a teenaged boy walked up and set up his video camera next to mine. As we both videotaped the action, I learned that he was also a fire buff, and was training as a paramedic. It was the start of a new friendship. He would share many adventures with me, especially as the newest and youngest crew member aboard my 1953 Ahrens-Fox pumper at parades and firefighter's musters all over California. A muster is an organized pumping demonstration, using antique fire engines to show the public what firefighting was like in the 'old days'.

Besides computers and radios, the video cassette recorder (VCR) is another valuable new tool inside the firehouse. A host of fire service instructional videotapes are available to train in the latest new firefighting techniques, and to learn how to use, repair, and maintain fire engines and firefighting equipment. Tapes even teach paramedic

Cab interior of F.D.N.Y. Engine 24, a Mack model CF pumper, shows the strategic placement of firefighter's air bottle, so it is easy to slip onto his shoulders and fix the mask in place. A gauge shows the firefighter how much air is left, so he can exit the fire building and grab a new air bottle before his air supply runs out. In case he forgets to check the gauge during the organized chaos of fighting the fire, a bell signals when the air in the bottle is running low. The firefighter then has time to exit before he starts breathing smoke instead of clean air. In days gone by, so-called "leather lung" firefighters did not worry about breathing smoke, because in all-wood buildings, the worst effect was a few hours of difficulty breathing. But fumes from today's synthetic building materials, such as Poly Vinyl Chloride (PVC), can permanently damage lungs and internal organs. The variety of chemicals used in today's manufacturing processes can produce fumes in burning factories that lead to cancer or some other terminal illness. So today's firefighters always make sure they have enough air supply while fighting a fire.

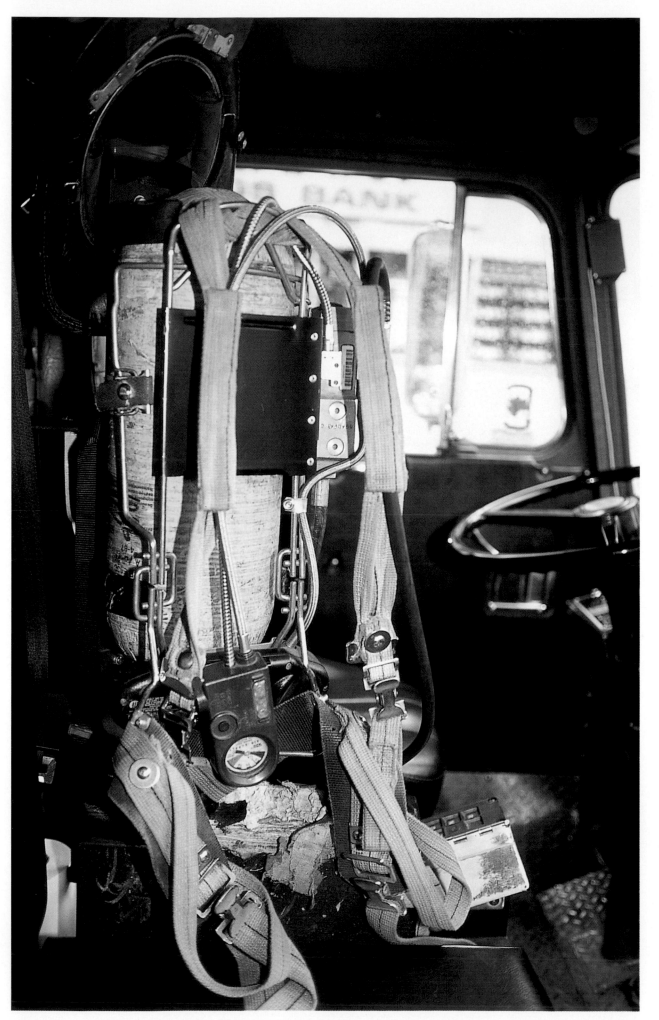

skills to use on victims of burns or smoke inhalation, car and truck accidents, train derailments, airplane crashes, drive-by shootings, and car jackings. And a tape of a recent fire could show firefighters would they did right (boosting their self-confidence), and more importantly, what they need to do better the next time.

Since the 1970s, firefighters have been more likely to need those paramedic skills than firefighting skills. Thanks to tougher building codes, more fire resistant building materials, fire sprinklers, and smoke detectors, firefighters now receive two or three times more medical-aid calls than they do fire calls. The firefighter/paramedic is now the average citizen's first line of medical help, providing everything from a bandage to heart defibrillation. Often, a paramedic call is for a situation so life threatening that, without this immediate, on the-scene medical aid, the patient would likely die while being transported to a hospital.

An important new tool in the late 1970s was the Hurst Jaws of Life. Using the tremendous pressure generated in an air compressor, its powerful jaws could quickly and efficiently pry open the door of a car or truck at an accident scene. A tiny pointed tool, about the size of a ballpoint pen, could smash away the windshield with a single tap (many firefighters find it quicker to smash away windshields using traditional fire axes). The powerful Jaws of Life could then quickly cut away the car's roof. Paramedics now had unobstructed access to the driver and passengers, to prevent emergency medical treatment for external and internal injuries.

Air compressors helped at the scene of derailed trains and overturned 18-wheel trucks, too. Starting in the 1980s, firefighters placed a lifting bag, made of a nearly-indestructible plastic fabric, under the overturned truck or rail car. As it filled from the air compressor, the bag tipped the truck or train car back onto its wheels, making firefighting and rescue efforts much easier.

The early American colonist, with his wooden rattle and leather bucket, would not recognize today's fire service at all. With the year 2000 fast approaching, who knows what new technologies will make today's sophisticated firefighting, rescue, and emergency medical equipment look antiquated?

The Hurst Jaws of Life, like this one used in New York City, uses the tremendous force of an air compressor to push two strong steel jaws apart. When wedged between door and body framework, these jaws can pry open even the most severely dented locked door. The Hurst tool can then be used like a giant pair of scissors to cut the roof pillars, so that firefighters can fold back the car roof. With doors removed and roof folded back, rescue crews can easily provide the driver and passengers with on-the-spot emergency medical treatment, and safely "extricate" the crash victims onto a stretcher, into an ambulance, and off to a hospital. With today's tough building codes and fire-resistant building materials, fire departments in the 1990s respond to far more emergency calls in homes and cars than they do to actual fires.

Chicago Fire Department looks after both O'Hare and Midway airports with a range of equipment, including the latest Oshkosh T-3000s, carrying 3,185 gallons of water and 420 gallons of foam. O'Hare has nine crash tenders, Midway three (numbers 6-5-2, 6-5-9, and 6-5-12).

Beverly Hills, CA, has the broad range of firefighting equipment and vehicles that would be expected, including Chevrolet Suburban four-wheel-drives, American-Lafranche tower ladders and pump engines.

Crew of F.D.N.Y. Engine 24 and Truck 5, in New York's Greenwich Village, model the latest fashions in firefighting turnout gear. The black coat is heavily insulated to protect against heat, water, and electricity. Coat and bunker pants are treated with Nomex, a man-made chemical compound that is highly resistant to flame. Even the most experienced firefighters are sometimes trapped, so yellow reflective stripes help firefighters find each other even in the densest smoke. Turnout boots have steel soles and heels, so firefighters walking blindly through thick smoke need not worry about stepping on anything sharp. Form-fitting work gloves allow maximum finger dexterity, while protecting hands from heat and flame. Two-way portable radios keep firefighters in constant communication with each other with the incident commander and pump operator, and if need be, even with the fire alarm dispatcher's officer. A simple triangular wood block, rubber-banded to a leather fire helmet, lets firefighters wedge doors open, so they are not trapped between the flames and smoke, and their means of exit. Note also flashlights attached to helmets.